CATCHING BREATH

Also available in the Bloomsbury Sigma series:

Sex on Earth by Jules Howard

p53: The Gene that Cracked the Cancer Code
by Sue Armstrong

Atoms Under the Floorboards by Chris Woodford

Spirals in Time by Helen Scales

Chilled by Tom Jackson

A is for Arsenic by Kathryn Harkup

Breaking the Chains of Gravity by Amy Shira Teitel

Suspicious Minds by Rob Brotherton

Herding Hemingway's Cats by Kat Arney

Electronic Dreams by Tom Lean

Sorting the Beef from the Bull by Richard Evershed
and Nicola Temple

Death on Earth by Jules Howard

The Tyrannosaur Chronicles by David Hone

Soccermatics by David Sumpter

Big Data by Timandra Harkness

Goldilocks and the Water Bears by Louisa Preston

Science and the City by Laurie Winkless

Bring Back the King by Helen Pilcher

Furry Logic by Matin Durrani and Liz Kalaugher

Built on Bones by Brenna Hassett

My European Family by Karin Bojs

4th Rock from the Sun by Nicky Jenner

Patient H69 by Vanessa Potter

CATCHING
BREATH

THE MAKING AND UNMAKING
OF TUBERCULOSIS

Kathryn Lougheed

BLOOMSBURY
sigma

Bloomsbury Sigma
An imprint of Bloomsbury Publishing Plc

50 Bedford Square
London
WC1B 3DP
UK

1385 Broadway
New York
NY 10018
USA

www.bloomsbury.com

BLOOMSBURY and the Diana logo are trademarks of Bloomsbury Publishing Plc

First published 2017

British Library Cataloguing-in-Publication data
A catalogue record for this book is available from the British Library.

Library of Congress Cataloguing-in-Publication data has been applied for.

ISBN (hardback) 978-1-4729-3033-0
ISBN (trade paperback) 978-1-4729-3034-7
ISBN (ebook) 978-1-4729-3036-1

2 4 6 8 10 9 7 5 3 1

Typeset by Deanta Global Publishing Services, Chennai, India
Printed and bound in Great Britain by CPI Group (UK) Ltd, Croydon CR0 4YY

Bloomsbury Sigma, Book Twenty-four

To find out more about our authors and books visit www.bloomsbury.com.
Here you will find extracts, author interviews, details of forthcoming
events and the option to sign up for our newsletters.

Contents

Introduction: I Caught TB from My Pet Cat   7

Chapter 1: Bringing the Dead Back to Life   13

Chapter 2: From Moo to Man and Back Again   29

Chapter 3: Didn't We Already Cure It?   49

Chapter 4: All That Glitters   65

Chapter 5: Thanks for the Memories   83

Chapter 6: The Human Universe   101

Chapter 7: Huber the Tuber's 20-Tuberculear Sleep   121

Chapter 8: Growing Fat on the Atkins Diet   141

Chapter 9: Killing the Unkillable   159

Chapter 10: The Drugs Don't Work   177

Chapter 11: A Barometer of Inequality   197

Chapter 12: Ratting Out the Missing 3 Million   217

Chapter 13: New Drugs for Bad Bugs   239

Epilogue: TB Continued   259

Acknowledgements   265

Index   267

I Caught TB from My Pet Cat

This is the story of one of the world's oldest diseases. Tuberculosis was a spectator at the birth of *Homo sapiens* and a passenger as these early humans took their first tentative steps outside the Cradle of Life to populate the planet. It was present as the first villagers threw down roots and it has witnessed entire civilisations decay into the ground. It marched with Roman armies and sailed the seven seas; walked with long-extinct mastodons and swirled in the wake of war and disaster. Sometimes a disease of poets and passion, more often a mass killer responsible for death on an unimaginable scale, tuberculosis has twisted through history alongside humankind, shaping populations and searing a trail through our folklore, art, literature and music. So it's perhaps a little surprisingly that I would start this book with a pet cat from Berkshire.

Poor Onyx, RIP, holds the dubious honour of being responsible for the first documented case of cat-to-human transmission of tuberculosis, or TB, after infecting his 19-year-old owner. Based on some of the news reports, you'd be forgiven for thinking Onyx was the black-cat harbinger of doom, foretelling the end of the world as we know it. 'Could you catch TB from YOUR tabby?' cries one headline in the UK's *Daily Mail*. That's right. Not just *any* cat, but YOUR cat. The beloved family member who's bonded with you through discarded mouse spleens and hair-caked bedsheets. Stop the press: Death personified (felinified?) has infiltrated close to 20 per cent of UK homes and right now he's coughing up a TB-infected hairball on your sofa. Should we be scared? Well, yes, damn right we should. That *Daily Mail* headline is truly awful.

Onyx was one of nine cats found to be infected with TB, at least five of whom died. One newspaper even published the names and photos of three of the victims – Onyx, Jasper and Mocha. I'm not sure it added much to the story but they were certainly very pretty cats. It's possible that they picked up their TB infections as the result of a run-in with a straycat who had taken on an infected badger (I have to admire its ambition). The *Express* felt it was necessary to explain that all the fighting and biting took place in or around Greenham Common, a location previously known for being the site of a radical feminist antinuclear protest. I keep trying to work out the relevance of this piece of information. As far as I know, TB is not an inevitable consequence of pacifism or assertiveness on the part of the female gender. Perhaps the newspaper was trying to ramp up the doomsday atmosphere of the story. Killer cats *and* feminists? Watch out, they could be coming to YOUR town.

The cats didn't actually kill anyone. They took their TB-laden open wounds back home and potentially exposed 39 of their human slaves. Two were to contract active TB, another two were asymptomatically infected. Everyone survived. So, not quite death on a nuclear-feminist scale. After a brief burst of media hype that lasted just days, everyone moved on to the next crisis to afflict Great Britain, and the cats were forgotten. But not by me. I keep looking at that photo of long-haired Mocha, with those alluring blue eyes that stare at something just out of shot. And I keep thinking, 'Is this the face of modern TB?' Has this mighty killer fallen so far that it's become tabloid scaremongering fodder that no one really takes all that seriously? When it comes to stories like this, part of me is pleased that people are talking about TB. Another part of me thinks that TB would no longer still be a problem if the actual humans who are dying evoked as much sympathy as a fluffy kitten. Most of me thinks that I have become rather cynical after 10 years in the TB research field.

I worked on TB first as a PhD student and then during two postdocs. So I have definitely exposed myself (academically, not literally) to more TB than my average fellow Brit. All of this means I have a slightly more realistic view of TB than the one portrayed by stories about killer cats. But what about everyone else? What does the mysterious 'general public' picture when it thinks about TB? A while back, I went through a phase of trying to bring the conversation around to TB whenever I met someone new. It wasn't an entirely successful endeavour from a making-friends point of view. It did, however, reveal that for many, TB is something that happens to people in different countries ... but that's as far as their interest went. Some of those whom I spoke to saw TB as a historical curiosity that had long ago been 'cured' in the UK; a number believed that 'it had come back', although only in migrants (no one mentioned cats). But overall, it wasn't really on anyone's radar as being a) scary, b) something that happens to the 'general public', or c) an interesting social conversation filler. TB, it seems, gives off this old-fashioned disease-of-the-past vibe. Like a smelly old blanket or antique dentures. If TB were a supervillain, it would be firing its PR company about now.

Consider this: if you had to take a guess at the biggest infectious disease killer on earth, what would you say? I suspect that a large number of people would go for malaria or HIV. So let's check out the stats. Malaria claims close to half a million lives a year; HIV/AIDS more than a million. TB, though? In 2015, there were 10.4 million new cases of TB around the world, of which half a million were drug-resistant and 1.2 million were associated with HIV. There were an estimated 1.4 million deaths in addition to another 0.4 million among people living with HIV. Be honest, you wouldn't have guessed 1.8 million TB deaths in just one year, right? I didn't think so. Hence my own small contribution towards rebranding TB as a modern monster rather than a mothballed relic of history. Political will to rid the world of TB is one of the big sticking points in turning all those millions into thousands, into hundreds, and into nothing. But the TB field hasn't always done the greatest job of making itself heard among all those voices trying to draw attention to their own causes. With so many other battles to be fought around the world, only some of them medical, TB isn't often top of anyone's agenda. I can't rank the world's problems by order of importance, but I do know that the TB situation is a sad indictment of our unequal world, where a curable disease continues to kill millions of people. Surely it's our ethical duty to ensure that everyone has access to life-saving medicines? Preferably before we reach the point where drug resistance sends us careening into an even worse world where TB is no longer curable.

We currently teeter at a pivot point in TB control. Since 1990, TB mortality has been slashed by nearly 50 per cent and new infections are reducing at a steady -1.5 per cent a year. Between 2010 and 2015, it's estimated that 49 million lives were saved through effective diagnosis and treatment. So, progress – but slow progress. At this current rate, it's going to take 150 years or thereabouts to reduce new cases yearly by 90 per cent and for them to finally drop below the 1 million mark. Because of this, in 2015, we saw the introduction of the End TB Strategy supported by the World Health Organization (WHO). They've set some ambitious targets to accelerate the fight against TB. By 2035, they want to see a 95 per cent reduction in TB deaths (compared with 2015), a 90 per cent reduction in incidence (that's new cases occurring per year) and no families faced with catastrophic costs due to TB. The first 'pillar' of the End TB Strategy is basically to get current treatments to everyone in need; the second is a 'let's do this thing' attitude from everyone involved; and the third is more science to improve the current intervention tools and how we use them.

It sounds easy, yet TB remains a huge problem. Six countries, in particular, account for 60 per cent of all new cases: India, Indonesia, China, Nigeria, Pakistan and South Africa. On a country-by-country basis, India has the most notified cases, accounting for 27 per cent of the global burden. It's followed by China, with 14 per cent. These countries, however, have massive populations, so absolute numbers don't always accurately represent where the biggest problems are. Throughout this book I often talk about rates of TB, usually measured in the number of cases per 100,000 people per year. When you look at these data, South East Asia and the Western Pacific are faring better than Africa. The African region reports 281 cases for every 100,000 people, which is more than double the global average. South Africa, in particular, has a whopping incidence of 834 cases per 100,000. There, no one publishes photos of pretty cats that have died of TB or refers to diseases of the past.

In high-incidence countries, TB is something that kills actual people. The majority of the time it's a disease of the lungs (pulmonary TB), but it can affect pretty much every part of the human body (extra-pulmonary TB). There's a game I like to play. I go to PubMed – a search engine of peer-reviewed literature – and type in 'tuberculosis+[any body part]'. Go on, give it a go. Barring an unusual knowledge of obscure human anatomy, I'm betting that you'll struggle to find a nook or nub that can't catch TB. For this reason, symptoms can be hard to pin down for extra-pulmonary TB. Take your pick from weight loss, malaise, night sweats, loss of appetite, weakness and fever. Pulmonary infections are characterised by these same general symptoms in addition to a chronic cough that sometimes brings up blood. The disease is spread via this coughing, which expels bacteria into the air to be breathed in by close contacts. Your family and best friends, to be specific. Not everyone who is infected will develop active disease. But of those who do, around half will die without treatment.

TB treatment is a bitch. In drug-sensitive cases, you're looking at two months of combination therapy with some seriously horrible antibiotics. No seven-day courses of amoxicillin here. We're talking isoniazid, rifampicin, pyrazinamide and ethambutol, given daily where possible. After two months, a patient is continued on 'just' isoniazid and rifampicin for a further four months. Things are much more complicated when you start getting into drug-resistant TB, where 20-month-plus treatment regimens are the norm. Treatment doesn't always work, even for drug-sensitive TB. A substantial number of patients will suffer a relapse after apparently being cured. Others will be reinfected. Unlike diseases such as chicken pox, a previous TB infection doesn't provide protective immunity against subsequent infections. In fact, having TB once makes you much more likely to catch it again. This fact always intrigues me. The idea that this disease has found its way round the human immune system; a consequence, I would hazard, of having had a lot of practice.

So, that's what this book is about. A whistle-stop tour through the past few million years to pick apart what has made TB 'TB'. A look at how evolution has shaped both the TB bacillus and the human response to the disease, entangling the two species over many millennia. Those are trees on the front cover. Lung trees. If the history of humankind is a tree, then TB is a vine, squeezing the tree's branches, cutting into the bark and altering the direction of growth. Untangling ourselves from this ancient foe is not a simple task. How do you kill something that's spent millions of generations finding ways not to be killed? That, at times, seems to know us better than we know ourselves? This book is the story of one of the world's oldest diseases – a scientific biography of sorts. Because if we can understand the makings of TB, then maybe we can find a way to unmake it.

CHAPTER ONE

Bringing the Dead Back to Life

Scientific diagrams? *Check*. Little boxes containing the remains of a dissected mummy? *Check*. Tapered candles fashioned from human body fat? *Wait ... what?*

It's 1825, and Dr Augustus Bozzi Granville has made a terrible error. He's about to present the findings of the first medical autopsy of an Egyptian mummy to the Royal Institution, and he's hit upon a way to add a little drama to his lecture. So he's had the room lit with what he believes to be the actual wax used to preserve the mummy. Only it's not wax. Granville has set fire to what is in fact 2,500-year-old mummy-tummy. Could there be anything better than sooty air tinged with burned human to add a little gravitas to a scientific lecture? No need to answer. The human horror-candles were actually Granville's second mistake in what was otherwise a well-considered and meticulous study. The first mistake was his conclusion that the mummy in question died from cancer. He was right in his identification of a large ovarian tumour, but wrong about it being the cause of death. I'll give you one guess at what really killed the mummy.

Dr Granville's story of scientific curiosity and killer microbes all started when the good doctor was called to treat one Sir Archibald Edmonstone. By this juncture in his adventure-filled life, Granville had set himself up as a renowned society doctor living in London. Driven, charismatic, somewhat pompous if his autobiography is an accurate representation. When Granville heard that Edmonstone had recently returned from Egypt in possession of both ill health and a mummy, the latter of which had set him back the grand sum of $4, he was overcome with professional nosiness. A day later, the dusty coffin was deposited in Granville's fashionable Savile Row dining room. There, over the course of six weeks, he carefully unwrapped and dissected the mummy in front of a host of up to a 'hundred scientists and literary characters', as he put it, who dropped by for, presumably, tea, biscuits and bandages. *This mummy is truly magnificent, old chap. Pass the jam roly-poly, will you?*

As an aside, I do have to wonder what his wife must have thought about the whole macabre spectacle. It's one thing to come home and discover that your husband has impulsively developed an interest in lawn tennis, or celebrity hair collecting, or laudanum (all acceptable nineteenth-century pastimes); but finding a dead ancient Egyptian laid out on your polished dining table surely must have come as a mild surprise. Granville does mention his family in his autobiography, and it's clear that he felt a considerable affection for the 'little ones', as he described his children. Yet I don't get a sense of their personalities through his writings, so, for all I know, the 'little ones' were permanently scarred as a result of Granville's mummy-themed escapades. In any case, regardless of his family's feelings, Granville set about his study with a child-like enthusiasm. The paper reporting his findings is quite a read. One thing that strikes me is how Granville's obvious excitement is tempered by a real respect for the mummy. Scientific papers of today often tend to be stiff and stilted in their language. It's all 'the experiment was performed' and 'from these data it is postulated that'. The language can be as dry as Granville's mummy, and the modern peer-review process would rapidly purge any whiff of personality from a paper. Granville, though, is not confined by such conventions. Oh no.

As I turn the pages, I begin to feel like he's narrating the paper to me in an Italian-accented, haughty baritone. He's pacing to and fro in his candle-lit dining room (burning wax, not fat), occasionally pausing to run a hand over his impressively bushy sideburns. 'I could not help experiencing a degree of enthusiasm, a portion of which, methought, I could impart to others,' he says, clearing his throat modestly. You have to remember that this is an adopted Englishman who lived during a rather repressed period in British history. Stiff upper lip and all that. But, in nineteenth-century terms,

Granville is leaping up and down and squealing with delight as he states, 'I determined, perfect and beautiful as it was, to make it the object of further research by subjecting it to the anatomical knife.' *Eeeek, it's a mummy. A MUMMY!* In my imagination, Granville lights a pipe and settles himself in a comfy armchair.

Of course, on the surface, his paper is all restraint and serious science (my flights of fantasy and Granville's flowery language/overuse of commas aside). He came at his study with the illustrious aim of unravelling the mysteries of embalming. How did the ancient Egyptians manage to preserve a body so well that, 25 centuries later, it was possible to comment on minutiae such as hair colour? Granville devotes several pages to meticulous descriptions of the bandaging process before he even gets to the mummy herself. *Herself.* The mummy was a mother in her 50s, of 'very considerable dimensions', with a perfectly dimensioned pelvis. Granville is very complementary when it comes to her pelvis. 'Not only are these the most perfect dimensions which a female pelvis can have, but they are precisely in the proportion which the longest diameter bears to the shortest, in the Venus of the Florentine Gallery.'

I start reading the paper as a scientific exercise, but by the time I'm halfway through, I'm trying to imagine who this woman was. The hieroglyphs on the mummy's coffin have now been translated. Irtyersenu: the lady of the house. That's all I have to go on. I come across a passage in Granville's paper that makes me pause: 'There appears to have existed no desire in the surviving relatives to preserve lineaments of a cherished friend.' Here he's describing how the bandaging process did little to retain her facial features despite such care being taken during the wrapping of the rest of her body. When I talk to John Taylor, an Egyptologist at the British Museum, he tells me that the mummy was so well preserved that you'd imagine the body was 'treated according to the best method at the time'. In other words, it can't have been a bargain-basement embalming package. So why were there no trinkets and jewels hidden within the bandages like an afterlife version of pass-the-parcel, as was often the case? Why was her face squashed, and why were most of her internal organs left within the body cavity?

The strange case of the organs all present and accounted for is particularly interesting. John explained that organs were usually removed through an incision in the body's abdomen or via an anal purge. So, was Irtyersenu's unusual method of preservation deliberate, or the result of sloppy embalmers wanting to get off work early? Whatever the explanation, these internal organs are the reason I'm including the mummy's story in this book. Without them, the circumstances of her death would have been lost to history. It's almost unbelievable when you think about it. Irtyersenu died, civilisations fell, the world changed. But when Granville peeled away those bandages, there was enough left of Irtyersenu to make a stab at diagnosing her cause of death, even if he didn't get it quite right. I'm reminded of my conversation with John Taylor when, on a whim, I asked him why he thinks people are so interested in mummies. 'Oh gosh, well that's the perennial question,' he said in a tone that suggested he'd been asked this before and didn't really want to have to answer. 'You've got these people who've kind of cheated death in a way,' he finally said. And that's a little how I feel about Irtyersenu. How many of us will still be around in thousands of years' time to talk to future generations about what death – and life – is like for us now?

Irtyersenu's true malady nearly went unsolved. Granville kept the mummy, or what was left of her after his experiments and misguided lighting arrangements, for many years after the autopsy. Then, in 1853, he donated it all to the British Museum. In his autobiography, Granville seems rather put out that instead of exhibiting it, the museum stored it away among their vast collection of specimens. Quite possibly the reason for this was Granville's inclusion of a few experimented-upon stillborn foetuses. Oh Augustus, why did you have to ruin everything with those foetuses? So his mummy parts remained lost in the archives of the British Museum for over

100 years until their rediscovery in the 1990s. With scientific techniques and knowledge happily having come some way since the nineteenth century, the remains of the mummy were re-examined. Scientists Mark Spigelman and Helen Donoghue are leaders in the field of detecting ancient pathogen DNA, in particular that belonging to the bacterium *Mycobacterium tuberculosis*. TB was rife in ancient Egypt, taking advantage of a sun-baked population who spent a lot of time indoors, presumably coughing and sneezing on each other. Mark and Helen believed that Irtyersenu's lungs showed evidence that she'd succumbed to TB rather than cancer. Proving it, though? That was where the ancient DNA came into its own.

When I talked to him, I pictured Mark Spigelman as a modern-day Granville, only without the mummy candles and the bloated ego. Trained as a doctor, Mark took a sabbatical at the age of 50 to pursue a BSc in Archaeology, planning to return to surgery straight after graduation. During this time, he started thinking about diseases in ancient remains. While any infection-causing bacteria would be long gone in a human skeleton, there was a chance they might have left behind traces of their DNA, much like a murderer fleeing the crime scene. What Mark needed was a technique capable of detecting the tiny amounts of DNA still present, a technique known as polymerase chain reaction (PCR), which, when Mark was making his first forays into archaeology, was still in its adolescence. 'I'd heard about the early work on polymerase chain reaction, PCR, and being an ignorant surgeon I thought maybe I could find the PCR in bones of some ancient bacteria,' Mark told me. He picked TB simply because it can lead to characteristic changes in a sufferer's bones, making it possible to identify human remains that might yield a positive PCR fingerprint. Mark got hold of some bones, found himself a student who was using PCR to diagnose TB in non-dead people, and his career as a surgeon was over. 'No one believed us at the time but it didn't matter, and several people tried to rubbish us too, but there are now hundreds of labs looking at various pathogens in ancient remains, so I think it virtually led to the foundation of a new science.' I scribbled 'Spigelology' in my notes, but Google informs me that this branch of science is more commonly known as Molecular Archaeology. Regardless, how many people can say they've founded a whole field?

The reason for the controversy associated with Mark Spigelman and his colleagues' work is that PCR involves amplifying a tiny, tiny amount of DNA millions of times. 'If you have to amplify something, there's a chance you'll get some contamination creeping in,' Helen Donoghue explained. Contamination can lead to false positives; a problem I've experienced many, many times in my own career. I've worked with some scientists who are so meticulous and precise that the very thought of contamination is alien to them, and others who contaminate everything in their path. We used to joke that, no matter what bacterial gene one colleague was trying to amplify, he'd always manage to end up with the human version, as if he was spitting in his reactions and adding a little of his own DNA into the mix. Helen's lab is clearly populated by scientists who most definitely do not spit into their experiments. But, despite all their care to avoid contamination, this potential issue has still got them into some tricky discussions with fellow scientists.

To further complicate matters, isolating DNA from Granville's mummy was particularly difficult due to the unique way that the mummy had been preserved, including the presence of some unidentifiable black oily substance that interfered with the PCR reaction. The results were not conclusive. But Helen and colleagues had another trick up their sleeves. *M. tuberculosis* is surrounded by a very thick, lipid-rich cell wall that not only goes some way towards protecting the bacterial DNA (meaning that TB is easier to diagnose than other ancient infections), but is also very stable in itself. Helen's collaborators tested for the presence of cell wall lipids in the mummy's lung tissue and thigh bone. 'It's an independent way of confirming the diagnosis,' she told me. 'You don't have to amplify the signal, and I think you can detect femtogram amounts.' I'm not sure exactly how much a femtogram is, only that it's a very, very small number. So the scientists were able to detect the ancient traces of *M. tuberculosis* using two independent techniques. And the mystery of Irtyersenu's death was solved. Mark and Helen's hunch had been right. Irtyersenu most likely died of TB that had spread throughout her body, making the poor woman very sick for some time before she finally found herself on the embalmer's table.

Intrigued by Irtyersenu's story, I decide to visit her in the British Museum. There's a certain solemnity about the museum, even during the school holidays when the place is crammed with 'little ones', as Granville would say. Walking through the expansive glass-roofed atrium, staring up at towering walls of white stone, I think about the necropolis at Thebes where Irtyersenu had intended to while away her death. Carved into the rocks, surrounded by sand-dusted mountains that just go on and on, the necropolis lay silent and forgotten until Sir Edmonstone's ilk started to pillage the tombs in their misguided curiosity. On the phone, I'd asked museum curator John Taylor about the nineteenth-century obsession with unwrapping mummies and whether it added anything to our scientific understanding of the ancient Egyptians. He sounded almost sad when he explained that, because Egyptology had still been in its infancy, the results of those experiments didn't fit into a good historical framework and, because the destruction they caused was irreparable, we can't go back and check what they found. 'The gain to knowledge compared to the destruction was not that great,' he said in a matter-of-fact way, wrapping up the conversation.

When I visit, there's an exhibition on at the museum where scientists have used modern techniques such as computerised tomography (CT) scanning to look inside the mummies' wrappings. 'The real aim of the exhibition is not just to focus on causes of death but to try to look at how people lived,' Taylor told me. 'What were their living conditions, life expectancy, diet, nutrition; what can we learn from human remains about these things? It's really important because that's exactly the kind of information we don't get from written sources from ancient Egypt. They don't tell you things like how long people lived, what they ate. They don't bring you close to the real people.' I wonder if the scientists performing this work felt anything like Granville as they virtually unwrapped the ancient bandages and took a peek into history while the mummies slept on, oblivious?

I find Granville's mummy on the museum's second floor. Irtyersenu is little more than a tray containing small pieces of what looks like old leather. Turns out it's hard to find the person in a few pieces of dry tissue and bone. There's something about museums that makes me feel a bit *Emperor's New Clothes* at times. It's like all the other people slowly and reverently filing past the display cabinets seem to spend more time looking at the artefacts than I do. Are they seeing something that I can't? There are times that I think this is probably a personal failing on my part – an inability to link the ancient artefact with real people who really lived. It's hard to believe that someone truly existed when they've been reduced to a tray of samples. Maybe this is also true of my scientific career. For me, *M. tuberculosis* has always been an intellectual puzzle to be solved, first; a killer disease, second. Sometimes I have to remind myself that, despite the nineteenth-century preoccupation with romanticising tuberculosis (I think the Victorians were a little death-obsessed, perhaps also explaining the 'mummy-mania' of this period in history), it's a gruesome disease. Like the devil, TB has gone by many names. Tuberculosis, consumption, white plague, schachepheth. Consumption is perhaps the most apt. The disease literally consumes people from the inside out. I'm not sure what part of that is in any way romantic. Yet, despite all of this, I've always had an affection of a sort for *M. tuberculosis*. Wait, wait, let me explain.

I remember meeting the Bloomsbury publisher when we first starting talking about this book, and he commented that my partner (a Formula One designer) and I were up there among the more excitingly employed couples he'd met. I think he shares this opinion with exactly no one. Because, usually, when I start talking about TB, people look at me like I've just admitted to being in love with a death-row serial killer and, by the way, we're getting married in the prison chapel next Wednesday. The point I'm trying to make here is that trying too hard to make a killer disease sound exciting often makes you seem a little psychopathic. It's not that I admire *M. tuberculosis* for its body count, but I am fascinated by how this bacterium has evolved to be so incredibly hard to kill. In my mind, there's a difference between wanting to share my love of bacteria and loving a disease that claims millions of lives. Bacteria have just always been my thing, right from when I was an odd little child. It's from this angle – why is *M. tuberculosis* such a good pathogen? – that I initially came to writing this book.

But Irtyersenu's story is as interesting because of the people involved as it is for the science. With this in mind, I started thinking about who the oldest known victim of TB might have been. Can science rewind time further than Irtyersenu? Written records take us back only 3,000 years to India, or 2,000 to China. More recently than this, the ancient Greeks referred to the disease as phthisis – a self-fulfilling word, in that it cannot be said without spitting on anyone in close proximity, greatly increasing the chance of infecting them. There are references to TB, or schachepheth, in the books of Deuteronomy and Leviticus, but in the end my search brings me full circle to Helen Donoghue and Mark Spigelman. This time, their work is taking us back further than the sprawling Egyptian cities, to the moment when humans first laid down their hunter-gatherer tools and settled into the farming life; long before Irtyersenu lived, died and lived again.

Atlit Yam is Israel's Atlantis. Through sunlit blue water, a semicircle of stone fingers taller than you or me juts out of the sandy ocean floor, barnacle-encrusted and slimy with algae. For around 9,000 years a layer of clay protected the village's ruins, shielding them from frequent storms and the ravages of time. Since Atlit Yam's discovery by the Israeli marine archaeologist Ehud Galili in 1984, excavations have revealed this site to be the archaeological equivalent of an unopened Lego Ultimate Collector's Millennium Falcon. It still has all the exciting little bits and pieces that it started out with, including a clay-brick wall, possibly erected to protect the village from floods during the winter months; houses; a workshop; a well; and two megalithic structures, one of which resembles a smaller version of England's Stonehenge. The arrangement of the 600kg (1,320lb) stones around a freshwater spring and the cup marks carved into the rocks suggest a ritual involving water. The megaliths were no doubt integral to keeping the community together by encouraging social interaction (religion? animal sacrifice? late-night parties?). Atlit Yam was, in fact, one of the very first places where enough people lived together for a community to be possible.

At some point, the hunter-gatherer founders of Atlit Yam must have thought to themselves, 'Hey, I can't be bothered to hunt tomorrow, so let's tie up a few cows and eat them later.' And one of the first-ever villages was born. Farming, fishing and animal husbandry were big in Atlit Yam. Goat, cattle, pig, gazelle, deer and fish skeletons were found during the excavations ... 6,000 or more fish skeletons, in fact. Combined with signs in human remains of an ear condition caused by regular exposure to cold water, we can be fairly sure that fishing played an important role in the village. When I was researching the history of Atlit Yam, I came across the idea that these piles of fish are evidence that the village was abandoned in a hurry as a tsunami thundered its way towards the shore. Could the drowning of Atlit Yam and other settlements have been the origins of the story of Noah's Ark, as has been suggested by some? I got quite excited, until further research revealed that the people behind this theory are perhaps not ... how do I say it? ... *basing their theories in reality*. It's a shame, as I really like the idea of Atlit Yam's destruction being the historical inspiration for the biblical story. But whether it was via a killer wave, climate change or divine intervention, Atlit Yam was lost beneath the waves and effectively frozen in time.

Not all of the villagers left before the submergence. Some were otherwise occupied with being already dead and buried in their marshy clay graves. The fact that the village's dead were interred close to or in the dwellings hints at some form of ancestor worship, so I have to wonder what the living felt about leaving them behind. I don't believe in ghosts in the literal sense, but there is something strangely haunting about these perfectly preserved bodies surviving all those thousands of years, refrigerated in their cool, dark graves until Ehud Galili and his scuba-attired archaeologists arrived to dig them up. Fifteen human skeletons were found – most in single graves, but one containing a 25-year-old woman buried with an infant roughly one year of age. The preservation was so good that Helen's team was able to diagnose TB in the mother and baby using both PCR and by detecting the more stable cell wall lipids.

The baby's skeleton was very small and, based on changes to its bones, was probably suffering from disseminated TB that had spread from the lungs throughout its body. Cases like this are usually the result of acquired neonatal TB, where an adult infects the infant shortly after birth. My new habit of attempting to picture what a long-dead individual must have been like does me no favours in this case. Listening back to the recording of my interview with Helen, it's possible to hear my own baby screeching happily in the background. I can't quite decide if it makes the story less sad knowing that the Atlit Yam mother and infant were buried together. The non-scientific parent in me feels slightly better for some irrational reason. The rest of me thinks that I shouldn't be sentimentalising deaths from many millennia ago. None of the skeletons were particularly healthy, after all. Most had very poor teeth, and some had issues with their spines. Perhaps ill health and childhood mortality were a day-to-day occurrence in Atlit Yam, but I'm not sure that makes it any better.

With untreated TB carrying a death rate of around 50 per cent, the deaths of both the mother and baby in close succession is, unfortunately, not especially noteworthy. What is special about these cases is that the preservation of the bacterial DNA was so good that Helen was able to determine the family line of the bacterium infecting the two people. We used to think that TB was an animal disease passed on to humans when we started living close to livestock so that we could eat them when peckish. Atlit Yam being one of the first villages showing evidence of domesticated cattle is therefore particularly significant here. Only, the strain infecting the mother and baby appeared to be more similar to modern human strains than the one that infects cows (we'll come back to this mystery in Chapter 3). The presence of cattle in the settlement is still relevant, even if they weren't the source of the infection. TB is a disease that thrives where humans are living in cramped, crowded conditions. The domestication of animals allowed settlements to grow in size and support a higher density of people by providing a major dietary component for the villagers. A study from 1988 estimated that TB requires a social network of 180 to 440 people in order to become endemic within a community. I doubt that earlier hunter-gatherers would have met that many people in their lifetimes, never mind socialised with them on a regular basis. TB, thanks to its ability to enter into a latent state analogous to hibernation, would have survived in hunter-gatherer populations but it wouldn't have thrived. It took the beginnings of urbanisation to finally bring out its best – or worst, depending on your perspective.

So, if the Atlit Yam tuberculosis victims are the oldest known cases, does that mean TB as a disease is 9,000 years old? Most definitely not. Searching for evidence of TB in ancient human remains is a little like ghost hunting. Most of the physical traces are long gone. All that's left behind is often a minuscule trace of ancient DNA, or femtograms of cell wall components, or characteristic bone lesions that could have been caused by TB but also could have been down to something else. It's difficult to tell. The oldest example of skeletal TB was suggested to be in a 500,000-year-old *Homo erectus* skull, with lesions and pits characteristic of tuberculous meningitis. Only, not everyone agrees. Concrete evidence in the form of DNA or cell wall lipids is limited by the availability of bodies well enough preserved to be able to detect these physical remnants of a TB infection. Cold, dry, no oxygen, no disturbances – Atlit Yam was the perfect combination of environmental conditions for the preservation of DNA. At the other end of the spectrum we have damp English graveyards. The oldest confirmed UK TB case from my home country dates to the Iron Age, thousands of years younger than the Atlit Yam victims. This unnamed man – who died somewhere in his 30s or 40s after what must have been a long and painful bout of TB – was dug up in a hamlet in Dorset. You know how you can tell a lot about the history of a tree by chopping it down and looking at the colour and width of its rings? Human teeth are much the same – the composition of strontium and oxygen can act as a fingerprint for where a person was born and raised. This can tell us about TB migration patterns and explain how the disease spread around the world. It was via this field of research that I happened across Charlotte Roberts, an English palaeopathologist.

Charlotte's work can involve examining human skeletons excavated when a new supermarket or school or shopping centre uncovers someone's long-forgotten grave. It's a legal requirement that evidence of archaeology at a building site must be thoroughly investigated, to decide whether any skeletons discovered should be reburied or curated at a museum for use in scientific research. To be entirely honest, I wanted to talk to Charlotte in the hope that I might uncover some juicy rivalry between her and Helen Donoghue. A paper from 2009 bearing Charlotte's name was highly critical of Helen's Atlit Yam paper, raising the old argument about contamination (PCR can detect such small amounts that a lab already dealing with *M. tuberculosis* could theoretically transfer some DNA onto the bones). Charlotte and co-workers also suggested that the destructive nature of ancient DNA analysis on bone was not justified in this case. Ouch. And do you remember that 500,000-year-old skull I mentioned above? Well, Charlotte was the author of the follow-up paper that provided the smack-down blow to the idea that its owner suffered from TB. She takes no scientific prisoners, it would

seem. But, when I talk to Charlotte, I find myself side-tracked
by her passion for reconstructing ancient lives and I forget all
about scientific rivalries. Instigating a scientific punch-up
between scientists feels distasteful when you start considering
the human remains involved as ... well ... humans.

Like Helen Donoghue and Mark Spigelman, Charlotte is
interested in ancient diseases, but comes at her work from the
direction of what the social implications are of disease on
people's lives. It's a perspective heavily influenced by her
former life as a nurse. 'It's a real holistic, whole-picture
approach rather than focusing on one thing. That's one of the
criticisms I have for biomolecular scientists who do this
fantastic DNA analysis – they're more interested in the sample
of bone you give them for the analysis than the bigger picture
of what the analysis means about that person or that
population.' Later in our conversation, she refers to scientists
in white coats treating skeletons as materials for analysis and,
by that point, I'm squirming in my seat. During all my years
in TB research, I met the grand total of zero TB patients. I
was most definitely a scientist in a white coat in this respect.
It was always all about the bacteria for me. But when I talk to
Charlotte, I do start to think about what I really want to
achieve here. My initial plan to shine a light on *M. tuberculosis*
and how clever the bacterium is suddenly seems a little, I
don't know, *cold*? Maybe I should be using this opportunity to
learn about how TB affects real people living real lives. 'I see
my job as looking after the dead rather than the living,'
Charlotte tells me. 'To try and put flesh on their bones, to
bring them back to life and to understand what kind of
challenges they had in their lives ... and from their disease
patterns, understand the living conditions they had.' It's a
person-first perspective that I never considered during my
own career.

Charlotte told me how the motto of her archaeological
society is 'Let the dead teach the living', and this is the reason
I've started this book with a chapter on TB in ancient remains.
The stories of Granville's mummy and Atlit Yam are about
more than simply diagnosing diseases in dead people. They're

about learning from the past so that we can plan for the future. The research of Helen Donoghue, or Mark Spigelman, or Charlotte Roberts, has the potential to teach us where TB came from and where it's going. Work such as theirs highlights just how long TB has been clinging on to humankind's coat-tails; learning, evolving, adapting. It's this long, long period of co-evolution that has turned TB into the disease it is today – and made it so difficult to keep under control. Mark Spigelman told me: 'There were many diseases, but tuberculosis was always there. The Black Death was an epidemic – it came and went. Tuberculosis was always there.' His theory is that TB has been with us since the beginning, hiding in the background, patiently waiting for a sign of vulnerability that it could exploit. He believes that the disease has spent millennia learning from us and taking advantage of our weaknesses. The Black Death that wiped out huge chunks of the European population during its peak? Maybe TB helped determine who would die during an epidemic, as those already suffering from one disease would be too weak to fight off another? I ask him if he thinks this is still happening today. 'What's a common cause of death from HIV?' he says, sounding perhaps a little more excited than the subject matter warrants. 'TB?' I offer. 'TB!' he cries.

But it's not just a one-way street. For all that *M. tuberculosis* learns from us, we also learn from the disease. Whenever any infection makes its way into a population, it very rapidly kills off those who are most susceptible. In this way, a disease selects for those people who are better equipped genetically to fight it off. TB has spent thousands of years training humankind via natural selection, gradually enriching the population with genes that help us to not die the moment we are infected. One step forward for *M. tuberculosis*, one step forward for humankind. It's a status quo that's been honed over millennia. Humanity and *M. tuberculosis* have effectively grown up together, constantly adapting to each other and neither ever winning the ongoing war. This book looks at the impact that both species have had on each other during their very, very long history. It's the story of a relationship that has stood the test of time, and of an infectious disease that has plagued humankind for millennia, leaving its traces in sunken Stone Age villages and Egyptian mummies and medieval skeletons dug up from a Sainsbury's carpark. And it's the story of a remarkably resilient species that has been shaped by our relationship with TB – genetically, socially and culturally, in more ways than you might imagine.

CHAPTER TWO

From Moo to Man and Back Again

Scuttling through the desert scrubland of the Bighorn Basin, Wyoming, United States, a pack rat darts into a concealed cave entrance and plummets 25m (85ft) to its death. Not that it's any consolation to the rat, but it finds itself in good company. Short-faced bears, collared lemmings, lions, cheetahs, camels, stilt-legged horses, bighorn sheep and the occasional woolly mammoth share its fate. Natural Trap Cave is a bell-shaped cavern full of bones, the oldest of which date back to the ice age. Over the ages, tens of thousands of animals have toppled into this deadly trap to create a layered trifle of mammalian history, with the pack rat forming the cherry on the top. The 2014 excavation team who witnessed the rat's unfortunate demise affectionately named him Packy Le Pew (because he started to smell after a few days), covered him with a little metal cage and made his corpse their mascot. Scientists, eh? Packy is a taphonomy experiment in action. Give him 10,000 years or so and his remains will have been covered over with sediment and perfectly preserved by the naturally refrigerated cavern, just like the other creatures that've lost their footing over the millennia.

One of the cave's victims was an ice-age bison, who sadly didn't earn himself a Twitter hashtag. But nearly 18,000 years later, he did have the dubious honour of becoming the world's oldest ancient DNA-verified tuberculosis sufferer. *Bison antiquus* looked like a modern-day bison, only better. Standing at over 2m (7ft) tall, weighing approximately 1.5t and upholstered in a thick woolly pelt, *B. antiquus* – like modern confectionery bars – is an example of how things often become smaller and less impressive with time. For over 10,000 years, *B. antiquus* strolled the plains of North America, nibbling on grass and occasionally being nibbled on by a sabre-toothed tiger (like a tiger, only *better*). Later, humans got in on the act and hunted the species to extinction, if the majority of experts are to be believed. But, for several millennia, millions of these unhuggable herbivores roamed the continent in thousand-strong herds. They were also all infected with *M. tuberculosis*, according to US palaeopathologist Bruce Rothschild. That makes *B. antiquus* a key piece in the puzzle of where TB came from and how it got to where it is today. Chapter 1 was all about the history of TB in human remains, but to understand the origins of TB and how it spread to the four corners of the world, we need to go back further in time. Back before the sun-baked cities of ancient Egypt; before the early farmers of Atlit Yam. Our trip into the past is heading for the very beginnings of TB, when humans were a distant speck on the evolutionary horizon and the ancestor to *M. tuberculosis* had yet to infect its first victim. On the way, though, we're going to stop off in the ice age, where the history of *M. tuberculosis* is interwoven with that of the non-human animals who once roamed this earth.

Bruce Rothschild is an expert on dead animal feet, among many other things. Specifically, bone lesions caused by a TB infection. As with human remains, animals infected with TB can experience similar bone damage as a result. Their foot bones in particular are a good place to look for these telltale eaten-away patches. Bruce has looked at a multitude of ancient bones from an array of species that date as far back as 100,000 years ago. Bovids including bison, longhorn sheep and musk oxen; rhinos, hairy pigs, camels and carnivores; even long-extinct mastodons. A mastodon, for the uninitiated, was like an elephant, only better: 3m (10ft) tall, really hairy, with tusks 5m (16ft) long. Of the 113 mastodon skeletons he investigated, half had TB bone damage. Based on these numbers, it's likely that nearly every late Ice Age mastodon in North America suffered from TB. These behemoths of the animal kingdom went extinct some 10,000 years ago, and Bruce thinks TB might have proved the final piece of really crappy luck that consigned the mastodon to a museum exhibit. Sure, climate change, thirst and hunting by 'pesky humans', as Bruce puts it, are the more mainstream suspects. But 'it's like plane crashes,' he tells me. 'Planes generally don't crash because of one problem. They crash because of a combination of problems.' It's an analogy that seems oddly misplaced when referring to what had to be one of Mother Nature's least flight-worthy creatures, but I get his point.

What I personally find surprising is how things seem to have come full circle when it comes to elephants and tuberculosis. The mastodon may have disappeared from North America's plains, taking with it the strain of TB that infected it; but it turns out that modern-day elephants happen to be rather susceptible to *M. tuberculosis*. A day's drive from Packy Le Pew's final resting place resides another Packy. This one is significantly larger and, at the time of writing, in better health. Other than his TB infection, of course. For a while, things weren't looking great for the oldest male of his species in North America. He ended up losing a whopping 635kg (1,400lb) in two weeks, and his 18-month treatment regimen – the same cocktail of drugs used to treat human TB – had to be recalibrated due to side effects. But he now appears to be responding well and, thankfully, hasn't shown any signs of active TB. Packy is just one of a number of captive elephants diagnosed with TB. Somewhere around 12 per cent of captive Asian elephants living in North America are thought to be infected, so the problem is reaching epidemic proportions. However, precise numbers aren't known, as it's notoriously difficult to diagnose TB in elephants: ever tried to fit a 5t (12,000lb) bull into an X-ray scanner? The problem first surfaced in 1996 when two Illinois circus elephants died from the disease and 12 of the employees working at the circus tested positive for exposure. A few years later, a number of animals at Los Angeles Zoo, including two elephants, were diagnosed with TB. Fifty-five employees tested positive, many of them having had contact with the elephants, although none of them developed active disease. It's likely that the elephants originally caught TB from an infected handler.

As with elephants and mastodons, the same goes for today's cattle and ancient bovids. In Rothschild's studies, he found TB in 19 per cent of the ancient bovids he examined. Some of his samples came from Natural Trap Cave, others from

Californian tar pits and others from archaeological sites all over North America and Europe. They represented a cross-section of prehistoric life. *All* of the different bovid species he looked at had TB among their number. The pigs and camels and big cats were a big fat no, but those cloven-hoofed ruminants? Yep, TB. What you have to remember here is that although only 19 per cent of the samples Rothschild looked at had the characteristic signs of TB, not every infected animal would have suffered damage to their foot bones. Just as not every human TB sufferer has Pott disease (TB of the spine), or every owner of a toddler currently has a banana wedged in their DVD player. If 19 per cent of Rothschild's bovids had the skeletal markers of TB, then a much larger proportion of the animals would have harboured an infection in their now long-lost lungs or other unfortunate body parts. Rothschild uses the word 'hyperdisease' – a pandemic infecting all bovids regardless of their location or age in history. Almost every early-cow, and almost every early-elephant. TB, it seems, was doing pretty well for itself long before human villages like Atlim Yam arrived on the scene.

You can't tell the story of TB without bovids. Specifically, domesticated cattle. *M. tuberculosis*, *Homo sapiens* and Bessie the cow: the biographies of all three are as interwoven as a leather basket. My own experience of cattle is limited to a childhood holiday in Wales when the cows literally came home and stuck their cold, wet noses through the window of our chalet. But this gaping hole of knowledge is nothing that Google can't fix. *Fun Cow Facts*. Close to 1 million results. Domesticated cattle are clearly not as boring as I'd originally thought. Local accents, 60kg (130lb) of saliva a day and an inability to descend a flight of stairs, like a Dalek. The best one, though? Apparently, cattle align themselves with magnetic fields while grazing. This cow compass was proposed back in 2008 by a group of researchers using Google Earth to spy on ruminants from above. Is it to prepare the herd for a rapid escape should a predator appear? Or an in-built cow GPS system similar to that used by birds or turtles during migrations? Or maybe a way to stay out of the sun's glare? A little more reading reveals that, sadly, no one knows, and the original paper is rather controversial among Google Earth researchers. No matter. It's nothing I can't scientifically test for myself with the help of one toddler and a local dairy herd.

They probably smell us coming. Cattle can detect aromas from 20km (6mi) away, you see. That's a better sense of smell than a shark, based on internet lore. And given that you're also 22 times more likely to be killed by a cow than a shark, I worry that we've underestimated the humble cow. After all, there are approximately 1.3 billion cattle in the world, around the same number as the human population of India or China. *They're everywhere.* A methane-powered invasion derived, I discover, from just 80 individual animals domesticated 10,000 years ago. Take any of the 1.3 billion modern cattle on earth and it will be able to trace its ancestry back to these pioneering 80 creatures. That's quite some population bottleneck. The scientists who discovered this proposed that all modern cattle are the result of a single domestication event in south-east Turkey, rather than multiple farming success stories. One talented farmer who succeeded in domesticating these ancient beasts, where everyone else failed (died). The ancestor of modern domestic cattle was the auroch, and it was as mean as hell. Rippling muscles, vicious 'come closer and I'll impale you' horns, a taste for human flesh (not really). But Julius Caesar said of aurochs that 'they spare neither man nor wild beast which they espied.' Which raises the question of why anyone would want to domesticate them in the first place.

My local herd, however, aren't giving off any murderous vibes. One of them looks at me, its jaw grinding rhythmically, and flicks up its tail to expel an impressively powerful jet of urine. 'Waz zat?' the toddler screeches, jumping up and down in delight. She knows exactly what it is. Perhaps, I think, this is the cattle equivalent of telling me I'm unwelcome. The herd is certainly not cooperating regarding the whole magnetic alignment thing. They're too busy eating and contributing to climate change. It doesn't matter, though, as the reason I'm really here is an old theory that it was cows that brought TB to the human race. This would make TB a zoonosis, or an animal disease transmitted to people. *M. tuberculosis* would be in good company were this the case, sharing its status as a zoonosis with many of the pathogens in circulation today. Rabies, West Nile virus, influenza, anthrax. All of them started out as diseases infecting other animals, jumping over to humans when the opportunity arose. Given the hyperdisease nature of TB among ancient bovids, it would make sense that maybe those 80 original aurochs were carriers. Then along came the Neolithic Revolution, in which people like those living in villages such as Atlit Yam started domesticating animals. Maybe an early farmer ate some tainted meat or milk, or maybe they were bravely living under the same roof as their killer aurochs. Either scenario could have led to the infection passing from cow to person, and the rest is history. It's a good theory, but don't get too attached.

The TB bacterium's modern-day family tree, at first glance, backs up the origin of TB as a disease of cattle. Today, the majority of human cases are caused by *M. tuberculosis*, but this isn't the only TB-causing species in circulation. *M. tuberculosis*'s dysfunctional cousins include species isolated from: voles (*Mycobacterium microti*), seals (*Mycobacterium pinnipedii*), goats (*Mycobacterium caprae*), antelopes (*Mycobacterium orygis*), mongooses (*Mycobacterium mungi*), dassies (the dassie bacillus), meerkats *(Mycobacterium suricatta)* and a recently discovered species that infects chimpanzees. And then there's *Mycobacterium bovis*, once believed to be the zoonosis that started it all. *M. bovis* is not particularly picky when it comes to its favourite flavour of host species. Cattle, deer, pigs, cats, foxes, possums, llamas, rodents – perhaps a better question is what species *isn't* susceptible to *M. bovis*. It is its taste for cattle that causes the biggest problem, though. At one point, around 40 per cent of UK cattle were infected with *M. bovis* in a situation reminiscent of ice-age bison, when entire herds carrying TB. Between 1912 and 1937, in England and Wales alone, about 65,000 people died from bovine TB. Before the advent of pasteurisation, the UK was seeing around 50,000 cases a year of TB in people who'd drunk contaminated milk from cows with TB mastitis. Today, the risk of transmission via raw milk persists in certain regions of the wood.

Attempts at eradicating *M. bovis* in livestock have met with variable success. Most of Europe, Australia and several Caribbean countries are now TB-free. Everywhere else, it's the presence of numerous wildlife reservoirs for the disease that cause all the problems. In the US, white-tailed deer occasionally infect livestock, although the rates are low; New Zealand has an issue with adorable possums spreading TB to deer and cattle herds; South Africa has TB in its buffalo and kudu; Spain has infectious wild boar; and in the UK, we have my self-appointed spirit-animal, the badger. This shy eater of earthworms and hedgehogs tends to avoid humans wherever possible but, when provoked, has a bite strong enough to crush bones. A *Daily Mail* story uses the lines 'like a horror movie', 'permanently scarred' and '48-hour rampage' when describing poor Michael Fitzgerald's run-in with one of these stripy beasts of doom. Another UK resident was reported as having his hand amputated after a badger chomped on his wrist. Yikes.

TB in British badgers was first described in 1971 during the investigation of an unexplained outbreak of TB in cattle. An association between outbreaks in herds and local infected badgers was soon proposed. In 2010, 215 animals from the Republic of Ireland were subjected to post-mortems to document the rate of TB among the population. Obvious TB lesions were present in 12 per cent, and when subclinical infections were included, the prevalence went up to 36 per cent. So, quite high. A 1998 study of 146 infected badgers revealed that half had lesions in their lungs, but that the infection could also be found in the kidneys (in 25 per cent of badgers), the lymph nodes (40 per cent), bite abscesses (14 per cent) and various other body parts from the brain to the bones. Many animals, at later stages of infection, develop advanced miliary disease, in which bacteria spread throughout the entire body. Unlike humans, badgers don't have to sneeze or cough on a cow to give it TB. Animals with miliary TB can shed vast numbers of bacteria. One study put the number of bacteria per millilitre of urine at 250,000, and 75,000 per gram of faeces. Perhaps a cow ingests some contaminated drinking water or hay, or perhaps the nosey creature happens upon a wee-soaked patch of grass and gives it a good sniff.

Over the last decade, bovine tuberculosis has cost the UK taxpayer more than £500 million, not to mention the cost to farmers, who can lose entire herds to the 'test and slaughter' policy in place to prevent further spread of the disease. Only we just can't seem to get a handle on it, and cattle TB is set to cost the UK another £1 billion over the next decade. Hence the controversial badger cull. The purpose of the cull is to remove enough of the badgers – 70 per cent, in fact – that transmission to cows is no longer a problem. But debate remains intense over whether culling is likely to be successful or whether it will actually make things worse. A randomised trial in 2003 did transiently reduce bovine TB in the local area (by around 50 per cent), but was associated with an increase in neighbouring areas. This effect is thought to be down to badgers roaming more widely after the cull, taking advantage of unoccupied territory and increasing the likelihood of spreading TB to cattle. The simplest answer here would be to kill more badgers over a wider area. But should our relentless hunger for meat and milk really outweigh the right of a beautiful albeit bone-crushing species to exist in peace? That's a topic for another book.

But what, you may ask, does any of this have to do with the origins of tuberculosis? Let's compare the organisms responsible for cow TB and human TB – *M. bovis* and *M. tuberculosis*. One infects a wide range of hosts (humans, cows and badgers included but by no means the only victims); one sticks primarily to people (although sometimes also other animals, such as elephants, if the opportunity presents itself). It seems kind of obvious that *M. bovis* would be the ancestral species from which *M. tuberculosis* evolved, right? It makes sense that human TB started off life as a zoonotic *M. bovis* infection passed from bovid to man. Then, as it persisted in the human population, it became lazy and gradually lost its ability to infect all those other species. A new strain branched off from the family tree and became *M. tuberculosis*. Only the theory all falls down for a number of reasons, including that ancient bison who met his end in Natural Trap Cave. Turns out he was not infected with *M. bovis* like modern cattle. Rothschild's study isolated ancient TB DNA from the damaged bison bones and sequenced PCR-amplified fragments. When they compared these small pieces of DNA to the same regions in modern TB species, the bison strain was 73 per cent similar to *M. bovis,* 77 per cent to *M. tuberculosis* and 82 per cent to another human strain called *Mycobacterium africanum.*

In 2002, a study authored by Roland Brosch demonstrated that 'which strain came first?' is an overly simplistic way of thinking about the evolution of *M. tuberculosis*. At first glance, *M. tuberculosis* and *M. bovis* look very similar. At the DNA sequence level, they are 99.9 per cent identical. But there are a number of regions of DNA present in *M. tuberculosis* that have been deleted from *M. bovis*. If you make a family tree of all the TB-causing species I mentioned earlier – the *M. tuberculosis* complex (MTBC), as they're known – then *M. bovis* isn't 'older' than *M. tuberculosis*. Instead, as the MTBC branch grew away from the family tree, it split into two. On one side, there was *M. tuberculosis*; on the other, *M. africanum*. It was from this *M. africanum* twig that all the non-primate-specific strains including *M. bovis* emerged. *M. tuberculosis* went on to do its own thing by diversifying into various lineages, which I'll come back to in a minute. What's important here is that the last common ancestor could not have been zoonotic *M. bovis*, as it emerged in parallel with *M. tuberculosis*, not before. The evidence, in fact, suggests that the disease, in some form or another, was around before humans even started to consider domesticating livestock.

If we go back to the roots of the family tree, they lie in the soil and date as far back as 1.7 billion years. Then, around 200 million years ago, the *Mycobacterium* genus emerged. This period of the earth's history was one of a worldwide temperature increase and the emergence of new forms of life, including the first mammals that would go on to become you and me. From the first primates that appeared somewhere around 75 million years ago, to the first of the great apes 15 million years ago, to the first of the *Homo* species 2.8 million years ago. *Homo habilis*, *Homo erectus*, Neanderthals and all the other stepping stones and dead-ends that arose on the way to *Homo sapiens*. During all this time, these early hominids would have been exposed to mycobacterial species. Today, there are a multitude of environmental mycobacteria out there, including some important model organisms used in the lab (*Mycobacterium smegmatis*, for example). During a conversation with Cristina Gutierrez, a scientist whose work has contributed to our understanding of the origins of TB, she pointed out a bottle of water that I'd recently been drinking from and told me that it was full of mycobacteria. She went on to explain how, when my daughter plays outside in the dirt, she's getting even more mycobacteria on her hands (and subsequently eating them). Mycobacteria are everywhere, she said. So for their entire evolutionary trajectory, our ancestors were exposed to members of this family of microbes. Along the way, they got really good at recognising them, possibly explaining why our immune systems react so strongly to *M. tuberculosis*. We've had millions of years of practice, and the memories of mycobacteria are there in our genes.

But how did what was originally a soil organism find its way to becoming one of the most deadly pathogens on earth? And what did this last common ancestor look like? A 2005 paper authored by Cristina Gutierrez proposed that, while today's MTBC is a relatively recent addition to the world, it's part of a much larger group of species that arose around 3 million years ago. The so-called smooth mycobacteria* include the mysterious *Mycobacterium canetti*. To date, only 60 or so *M. canetti* isolates have been discovered, mostly from opportunist infections of patients confined to the Horn of

---

* When grown on an agar plate, *M. tuberculosis* looks like crusty old droplets of spilled porridge – 'rough colonies', to use the scientific description. But when *M. canetti* was first isolated from a 20-year-old French farmer with TB, it looked more like drips of yoghurt. At the time, it was thought to be a smooth subspecies of *M. tuberculosis*.

Africa. Cristina explained how, at the time of her work on
M. canetti, she was based at the Institut Pasteur. It was here
that Georges Canetti had isolated the first strains of
M. canetti in 1969. Cristina took a cross-section of the strains
available to her, including Georges Canetti's samples and a
number of more recent isolates, and compared their DNA
sequences. More accurately, she had one of her students
compare the sequences. The results were so unexpected that
she had the student repeat everything, not believing that
there hadn't been some mixing up of tubes or mislabelling.
When the results came back the same, Cristina repeated all
the work herself. In the end, she had no choice but to believe
the data and start to consider what it all meant.

In her paper, she proposes that the small number of
M. canetti species in circulation today are so different from
each other and from the MTBC that they would have needed
between 2.6 and 2.8 million years to diversify to this extent
(this number is only an estimate, as dating the age of species
is notoriously difficult. I'll come back to that in a minute but,
for now, let's just say that the *M. canetti* family line goes back
a long way). If you want to imagine the MTBC and *M. canetti*
families as evolutionary trees, the MTBC's is a bonsai
compared to *M. canetti*'s Great Banyan. An evolutionary tree
that big doesn't grow in mere thousands of years. The 60
isolates from human infections likely represent just the fringes
of what is – or was – a much larger family. Somewhere out
there, there's surely an environmental reservoir for *M. canetti*.
This could make *M. canetti* the missing link between the
environmental mycobacteria lurking in our water bottles and
back gardens, and the pathogenic MTBC. A 2013 paper from
Roland Brosch's lab took this idea further and reported that
the smooth mycobacteria species are less persistent (good at
surviving for long periods of time in the host) and less virulent
(good at causing disease) than *M. tuberculosis*. This fits with
the idea that the last common ancestor of the MTBC started
out existence as an *M. canetti*-like species, before gradually
acquiring the persistence and virulence mechanisms that
make it what it is today.

What I like best about this theory is that it means TB could potentially have been a disease of early hominids long before *Homo sapiens* arose, making it older than the plague, typhoid fever or malaria. A 500,000-year-old *Homo erectus* skull discovered in Turkey provides some (albeit controversial) support for this. The scientists who studied the skull identified a number of smooth pits that they believe are indicative of a TB meningitis infection. Could it be that, long before the MTBC arose, *M. canetti*-like species were already killing our human ancestors? Was it a similar species to that which roamed the plains alongside ancient bovids? And was TB a contagious disease among hominids as it is today, or an occasional event similar to modern *M. canetti* infections? Unfortunately, it's nearly impossible to know. Bacteria don't leave behind fossils, and ancient DNA has a finite shelf life. Studies of taxonomy can suffer from a kind of survivors' bias as we only have the ones that make it left to study. In reality, species rise and fall and undergo population bottlenecks in which entire branches of the family tree are pruned from existence. Even the *M. canetti* strains we have to play with today are distant echoes of their own precursors. Attempting to reconstruct the last common ancestor from more recent strains is a little like trying to predict what someone's great-great-[insert more greats]-grandma would have been like based on the characteristics of her distant descendants. Not only has the DNA been diluted over the years, but you have to take into account how the world has changed since they were alive and how their environment might have influenced their personality, appearance and genes on an epigenetic level.

What we do know more clearly is that, somewhere between 6,000 and 70,000 years ago (roughly speaking!), the MTBC emerged. In a 2013 *Nature* paper, Sébastien Gagneux tried to reconstruct its evolutionary history. The problem with this line of research is that the evolutionary stopwatch doesn't keep very good time. Working out when the last common ancestor of the MTBC lived uses what's known as a molecular clock, which involves counting the mutations that have accumulated between related genomes. In simple terms, the more differences between two strains, the more time has elapsed since they diverged along different branches of the evolutionary tree. This relies on the mutation rate being constant between species and over time. It's not. Compare modern *M. tuberculosis* lineages, and the mutation rate can differ by up to 10-fold. And that's not even taking into account the differences between growing something in a lab and the conditions it would be exposed to in a real-life host. The growth and division of *M. tuberculosis* within the human body isn't constant. During an active infection, many of the bacteria are dividing rapidly and accumulating accidental mutations relatively quickly. But during a latent infection, the majority of the bacteria aren't doing very much at all. On top of that, mutations are not only accumulated over time but also lost, and it's obviously quite difficult to account for this when calibrating the clock.

Sébastien explained how, for *M. tuberculosis*, a few thousand years is a mere blip in its much longer history. Relying on mutation rate to date the MTBC just doesn't work very well. But when Sébastien's team created a tree using 250 modern strains from all around the world, they noticed something interesting. It looked remarkably similar to the pattern of human migrations that carried *Homo sapiens* Out of Africa and around the world. As the human race branched off and diversified, so did the MTBC. As human populations boomed during the Neolithic Revolution, so did the MTBC. So, Sébastien used the known dates of human migrations to take a guess at the age of *M. tuberculosis*. His work put the MTBC complex at 70,000 years old. This was long before the point at which humans started to domesticate livestock, again suggesting that the modern MTBC had a human rather than animal origin. It fits with the observation that both *M. tuberculosis* and *M. africanum* infect humans – as does *M. canetti* when the mood takes it. So it follows that the last common ancestor to the MTBC was also already a human-adapted strain.

I love the implications of Sébastien's theory, even if not everyone agrees with his dates. Even Sébastien is quick to tell me that he may be wrong and that his work relies on a whole lot of assumptions. In time, new views will evolve as new tools and data emerge. This whole chapter, in fact, is an attempt to construct a story from the sparse scientific evidence. One thing scientists do agree on, though, is that TB came out of Africa. Sébastien's paper proposed that the waves of humans leaving the savannas of East Africa to colonise the world took TB with them, *M. tuberculosis* hitching a ride with these early explorers. Many scientists believe that the human Out of Africa migrations occurred in two main waves, with climate change likely influencing the direction that the populations of early humans took. Two waves would certainly fit with the predicted ages and distributions of the modern *M. tuberculosis* lineages. You see, despite the global movement of humans made possible by air travel, particular lineages of *M. tuberculosis* still more or less stick to their original territories. Lineage 1 has made its home in the Indian Ocean region of the world, Lineage 2 in East Asia and Lineage 3 in Central Asia, while Lineages 5, 6 and 7 are found only in parts of Africa. Lineage 4 started off life as the European lineage, although it is now found throughout Europe, Africa and the Americas. Let's follow the potential paths that all these different lineages took as they left Africa.

To begin with, an initial wave of maybe as few as 150 Africans crossed the Bab-el-Mandeb Strait of the Red Sea in tiny rafts to reach the greener grass of the Arabian Peninsula. These fish-loving migrants then took a leisurely stroll along the coast of the Indian Ocean to colonise the West Pacific region and, eventually, Australia. Today, genetic traces of this population can be found in the Australian Aborigines, Papuans and Melanesians. The exact timings of this migration are up for debate, so let's just say it all happened sometime between 60,000 and 125,000 years ago and leave the arguments to the human geneticists. Some of these early humans were infected with *M. tuberculosis*, and these pioneering bacteria went on to become Lineage 1, or the Indo-Oceanic Lineage. It's now a relatively rare lineage, spread thinly along the Indian Ocean tracking the original wave of migration from the east coast of

Africa to India, Bangladesh, Vietnam and all the way to the Philippines. It's one of the ancient lineages, likely retaining more of its distant hunter-gatherer roots than some of the other branches of the family tree.

The more progressive, open-to-change lineages emerged from the strains infecting a second wave of human migrants. This population followed the Nile, heading north through the Sinai to colonise East and Central Asia. Along the way, they dropped off some of their number to traipse in the opposite direction towards Europe. As someone with mixed English and Canadian ancestry, my distant ancestors were likely among those who took this cooler route. Predictably, each group took *M. tuberculosis* with them. For the East Asian settlers, it was Lineage 2 (also known as the East Asian Lineage); Central Asian settlers had Lineage 3 (Central Asian Lineage); and my European ancestors brought Lineage 4 (Euro-American Lineage) along for the ride. These three lineages are known as modern strains. They're the bad guys who are causing most of the world's TB problems nowadays. Lineage 2 spawned what is known as the Beijing Lineage – basically, *M. tuberculosis* on steroids. We'll be back to him later.

Conversely, Lineages 1, 5, 6 and 7 are known as the ancient strains. Lineage 1 is the only ancient strain that made it Out of Africa, as I mentioned above. Lineages 5 (West African 1) and 6 (West African 2) evolved alongside populations of humans who remained in Africa, while Lineage 7 appears to have arisen after the proposed Out of Africa migration of the MTBC but is now found only in Ethiopia. It's possible that its early hosts were a population who left Africa but then returned at a later date. Perhaps they were homesick, who knows? Compared to modern lineages, the ancient strains are less successful in terms of their distribution around the world. Interestingly, it seems likely that higher virulence and a shorter latent period in modern strains better adapts them to life in the urbanised world. Could it be that the ancient lineages are a window back in time, towards the strains that would have infected ancient hunter-gatherers? A disease that rapidly kills its hosts needs a constant supply of new people to infect, otherwise it will burn itself out.

Back when we lived in small groups, foraging and hunting in a confined home range and rarely mixing with other communities, diseases ran the risk of killing too many of their hosts and finding themselves with nowhere to go. Before the Neolithic Revolution, there just weren't enough people living in close proximity to each other for a disease with no known animal or environmental reservoir and as high a mortality rate as TB to thrive. So it came up with a solution: latent infection.

During its long history with humankind, *M. tuberculosis* has learned to be patient. It can happily survive for sometimes decades in its human host without causing any outward signs of disease, only to reactivate when the host's immune system is no longer strong enough to fight it off. The 10 per cent (based on today's numbers) of individuals who develop active TB ensure that the disease spreads to others. The 90 per cent who harbour a latent infection are the bacterium's version of cryogenic preservation. This ability to form a latent infection is, even today, one of *M. tuberculosis*'s biggest strengths, but the lineages do seem to vary in just how patient they're willing to be. The ancient lineages can be viewed as a throwback to the days of hunter-gatherers, when slow and steady won the race. The modern strains, in comparison, appear to have adapted to today's world, with its overcrowding and high population densities. They've shed a little of their patience to take advantage of a steady supply of new hosts.

Of the modern strains, Lineage 4 is the most successful in terms of distribution. It is no longer confined to Europe but has spread across the world – explaining why the lineage is often called the Euro-American Lineage. But wait a second. If you've been paying attention you might have noticed that I didn't get to explaining how TB first came to the Americas. Surely the Central Asian populations who became the Paleo-Indians when they migrated across the Bering Strait should have taken their lineage with them? Shouldn't it be the Central Asian-American Lineage? Nope, it's not quite so simple. When you look at the predominant lineage in the

Americas, it's of European origin, not Asian. America, unsurprisingly, makes everything both a little more interesting and complicated at the same time. In the past, lots of people have hypothesised that it was the European settlers who brought TB to the Americas, explaining why the European strain predominates there. Part of this hypothesis was based on the unhappy knowledge that the Native American populations were decimated by the disease soon after Columbus and his ilk arrived. This has been used to suggest that the native populations were TB-naive, with no way of fighting off an infection their immune systems had never seen. But the evidence actually suggests that the Native Americans succumbed to TB thanks to their forced relocation into overcrowded, half-starved, disease-ridden reservation settlements. So, it's true that European settlers did kill the indigenous population with TB, but not in the way we originally thought.

In any case, the theory that TB arrived in the Americas with Columbus doesn't explain the Temple of the Seven Dolls – a Mayan archaeological site in the Mexican state of Yucatán. The temple is a squat, square building perched on top of some steep stone steps. It's an ugly little thing, reminding me of something a posh Englishman with far too much money might build in his sprawling Cotswold estate. *Won't you join me in my mock-Mayan temple for a spot of Pimm's, dear boy?* It gets its highly exciting-sounding name from the seven small terracotta effigies discovered beneath the floor. They're creepy little things, seemingly modelled after diseased individuals. One of them has what is medically known as an acutely kyphotic upper back. It's a deformity of the upper spine that gives a sufferer a dorsal fin-like hump and is most commonly observed with Pott disease – TB of the vertebrae.

The statues date from the period 900–1521 AD. If the curved spine represents a pre-Columbian Mayan, then perhaps TB existed in Mayan society long before European settlers made their way down into what is now Mexico. It's circumstantial evidence, though. More concrete proof that TB in the Americas predated Columbus came from ancient

DNA analysis of the remains of a 1,000-year-old woman who once lived near the Chiribaya Alta site in South America. Johannes Krause's study looked at 68 Peruvian mummies, all of which had skeletal damage indicative of TB, and found the remnants of mycobacterial DNA in three. The mummies all died somewhere around 750–1350 AD, long before Columbus set sail.

The scientists went on to use a molecular clock to estimate that the strain was less than 6,000 years old (of course, Krause's dates are subject to the same weaknesses as Sébastien Gagneux's). But if this figure is accurate, it would mean that the mummies' strain of TB couldn't have arrived with Native American populations. The 1,000km (600mi)-wide Bering Land Bridge connecting Asia to Alaska disappeared more than 10,000 years ago. So if Krause's TB strain couldn't have come to North America by land, it had to have come by sea prior to Columbus's voyages in the fifteenth century.

The final piece of the puzzle came from comparing the Peruvian strains to 40 other strains of TB bacteria. They were not like modern human-adapted forms. Instead, their closest relative was *M. pinnipedii* – the seal and sea lion strain of TB. Could it be that seal hunters caught TB from their prey and these strains went on to seed the Americas with TB, or is it more likely that these were isolated zoonotic cases? To me, the latter seems more likely, but this is still a brilliant example of how the history of *M. tuberculosis* is closely intertwined with both that of humankind and that of the world's other animals. I personally suspect that at least some TB did cross into the Bering Strait with the early settlers travelling from Asia. The East Asian Lineage (Lineage 2) can be found in the Americas today; it's just that the Euro-American Lineage (Lineage 4) now makes up 80 per cent of the cases. Like European settlers of the human variety, Lineage 4 may simply have risen to outnumber the natives, gradually replacing them over time.

No matter how exactly it got there, *M. tuberculosis* has made itself at home across the globe. It's come a long way from its lowly roots in the environment, gradually evolving and diverging … first into an *M. canetti*-like species that made the leap from a peace-loving existence to a pathogen … then into a strain capable of transmitting between its first hosts, whoever or whatever they were … on to a TB-causing species that may well have afflicted early hominids … and then to the last common ancestor of the MTBC. From there, the family spread around the entire world and gradually adapted to life alongside a huge range of species, strains rising and falling with both hominid populations and with the animals who've suffered from the disease during its long history. Some strains have been lost to time, such as those infecting early bovids – all we have to go on today are the scars they left behind. Others have flourished as the world around them changed in their favour. Along the way, *M. tuberculosis* has hitched a ride with numerous species, finding its way to nearly every human population in every corner of the world. From ancient humans, to fearsome bison and ice-age mastodons, to the emergence of *Homo sapiens* in the Cradle of Life, to modern-day cattle herds and Packy the Oregon Zoo poster elephant. As humans progressed onwards and upwards from the Neolithic Revolution, the bacterium's patience finally paid off. We handed TB everything it needed to become a global killer – urbanisation and, with it, a steady supply of new hosts.

CHAPTER THREE

Didn't We Already Cure It?

It's like the opening sequence of a murder mystery. Three 10-year-old boys playing in a rain-battered gravel pit, sliding down the steep banks, stone dust beneath their nails and pebbles in their shoes. It's all fun and laughter until one of them dislodges two human skulls that bounce down the slope and come to rest at the bottom. Staring with those empty sockets as the children scream in delighted terror. Of course, no one believes the boys when they recount their tale. At least not until one of them brings a skull home to meet his mother. It was 1990 and the town was Griswold, Connecticut. A few years previously, local serial killer Michael Ross had been sentenced to death, so immediately thoughts turned to undiscovered murder victims. But the bones were flaking and showing signs of decomposition: mud engrained into the tiny holes on their surfaces. These remains were old. So the medical examiner put down his tools, the police removed their crime-scene tape and everyone breathed a sigh of relief.

Old bones lie within the domain of the state archaeologist who, at the time, was a man called Nicholas Bellantoni. Arriving at the site, he quickly noticed six dark stains marring the horizontal strata in the rock. The excavation of the gravel pit had cut straight through an old graveyard, and these were burial shafts. Further investigations revealed that it was an ordinary Christian burial site belonging to the Walton family, who'd lived in the area from 1692 until the early 1800s. Much of a state archaeologist's work involves identifying remains unearthed by modern developers and ensuring they are reburied somewhere safe from diggers and 10-year-old boys. The Waltons should have been a simple case, but then Bellantoni's team uncovered a strange stone crypt. It contained a black-painted coffin with metal tacks hammered into the wood. From murder mystery, to the History Channel, and back to sinister thriller. In a particularly creepy turn of events, the tacks spelt out Bellantoni's own initials and the number 13 – the date of his birth. 'This guy's got my initials, he dies on my birthday, and I'm digging him up,' he says in one of the many talks he has given on this case. When the audience's laughter dies down, he continues with: 'The good news is when we examined the skeleton, it wasn't a short dark Italian and I felt a little better about it.'

So the remains weren't those of a time-travelling future version of Bellantoni. In fact, it was a 13-year-old boy buried alongside two graves containing his parents. That's the thing about eighteenth-century burial sites – most of the graves belong to children. Approximately one in three individuals didn't make it past 15 years of age during this period of history. A whole host of horrible diseases circulated in populations – the plague, smallpox, measles, influenza, dysentery. There was no germ theory or antibiotics; no one understood the benefits of hand washing or that disease could be spread by coughing or sneezing. The darkest part of this story is still to come, though. When Bellantoni peered inside the father's crypt from the back of the headboard, he came face to face with a skull that, instead of lying in the traditional position that a head usually occupies, was turned to Look Right at Him. Not only that, but the skeleton's leg bones were crossed over his chest pirate-style, and it looked like someone might have been rooting around inside the rib cage. All of this had happened three to five years after death. To surmise, someone had dug up the body, rearranged his bones, decapitated him and resealed the crypt. So much for resting in peace. A closer examination revealed TB lesions on the man's ribs, giving a clue as to his cause of death but not the reason for his bizarre exhumation and reburial.

'Have you heard of the Jewett City vampires?' a colleague asked. Because, as it happens, Bellantoni's graves showed some striking resemblances to a number of historical reports of corpses being similarly exhumed shortly after death. One of the stories goes that, back in 1845, Lemuel Ray had died of what we now know was TB aged 24, followed by his father four years later and a sibling two years after that. Three years passed and yet another brother – Henry – also began to sicken. The family was desperate to ensure that he didn't join his relatives in the family graveyard. Because the nature of TB as a contagious disease was not understood, there was talk that maybe Henry's father or brothers were feeding on young Henry during the night. So the locals unearthed the family and, discovering blood still in Lemuel's heart, took it as a sign that he was among the undead. They burned the heart, rearranged his bones, and buried him again. And Henry lived … briefly. And then he died. One of the most famous of the New England vampire stories has a similarly bleak ending. A local woman called Mercy Brown was to die from TB in 1892, alongside her mother and sister. Soon after, her little brother Edwin also fell sick. It's possible that the family turned to doctors and the church and anyone else who might be able to help. No one could. Then, at some point, the locals came upon this (at the time) widespread vampiric folk legend focused around the transmission of TB. Could it be that one of the recently deceased women was harming poor Edwin from the grave? They exhumed Mercy's body to check for signs that she was undead, and finding blood in her heart rearranged her corpse just like the other cases. Her heart was burned and Edwin drank the ashes in a remedial potion. He was to die just a few months later.

You have to remember that this was little more than a hundred years ago, yet the behaviour of these people seems almost unbelievable by modern standards. What was it about TB that captured the imaginations of the time and pushed people towards taking such extraordinary measures to protect themselves? I couldn't work it out, so, one dark and thundery night, with the fingers of a tree scratching against my window, I gave Nicholas Bellantoni a call to ask him about the vampires. Thanks to the time difference, it was a toasty Connecticut afternoon for him and our chat put an end to my visions of creepy graveyards and blood-sucking monsters. He explained how mistaking a TB patient for the victim of vampirism isn't that strange a conclusion to come to. 'You had tuberculosis victims sitting at the table, coughing and coughing, and no one's covering their mouth,' he said. 'They had no clue how this was being transmitted, so it's going through families epidemically. Those same victims might be sleeping in crowded bedrooms with maybe four or five other brothers and sisters, again coughing all night.' Based on the common causes of death at the time, every family would have buried at least some of their number as a result of TB. It really was everywhere, but also unpredictable. There could be days or months or years between victims. No one had a clue as to why the disease kept reappearing again and again within the same families – an understanding of latent infection remained a long way off. With no explanation or hope forthcoming from the living world, perhaps it's not so unbelievable that people turned their attentions to the restless dead.

On top of this, TB sufferers would often slowly waste away as if the life was literally being sucked out of them – hence the historical name for TB: 'consumption'. You have to consider how, when the bodies of loved ones were exhumed to check for evidence of vampirism, family members would often have been met with a person who looked nothing like the one who'd been buried. Gases would have caused the body to bloat, the skin would have shrunk back from the nail beds and hair follicles, and blood may have seeped into the mouth. To a pre-twentieth-century person, this could easily have been taken as evidence that the greedy corpse had been waking at night to feast upon the living. When I first read about the case, I was imagining something involving torch-lit trips to the cemetery and wild-eyed villagers. Twenty-six years of media interest, though, seems to have robbed Nicholas Bellantoni of all sensationalist leanings, if he ever had any. He described the theory as the dead being mischievous rather than something 'evil'. And he believes that the actions of the living came from a place of love and fear. 'To protect your family you had to go back into the graves and find out who was undead, and make sure you did the necessary procedures to make them dead-dead,' he told me. It all seemed quite matter-of-fact, the way he put it. He believes that were a disease such as Zika or Ebola to run rampant beyond what the scientific and medical community could handle, then you or I would probably resort to trying anything – no matter how ghoulish – to save our loved ones. When I imagine my own family dying one by one from a mystery illness with no cure, I can see his point.

The story of the New England vampires, for me, represents so much of the historical relationship between TB and humans. The fear which this disease must have evoked in communities; the helplessness as fathers, mothers and children sickened and died. At its peak, TB was responsible for a quarter of all deaths, often claiming those in the prime of their lives. This was one of the gifts of the Neolithic Revolution, which saw humans begin to live in larger settlements where disease ran rampant. I've read that the history of human death can be divided into three ages – the Age of Disaster, the Age of Disease and the Age of Decay. Prior to the Neolithic Revolution, during the Age of Disaster, people were likely to die of starvation, or accidents involving wild pigs, or being clubbed on the head by a rival. Disease existed, of course. But the crowd diseases we know and hate today were yet to come into their own. At the other end of the timescale, some of us have now tipped over into the Age of Decay, dying from diseases resulting from our own bodies failing us. Many, though, are still living in the Age of Disease. It's a period of time that encompasses more than 10,000 years of human history, and it all started with the advent of farming.

This dramatic change in the human way of life had a massive effect on the world's population, allowing it to grow exponentially. As a result, transmissible disease took over from violence as the top way to die. New plagues arose and old ones metamorphosed into something infinitely worse. The more people living together in one place, the easier it was for infections to spread: from person to person, via animals, or through contact with human waste. On top of that, a limited diet, the beginnings of social inequalities and

occupations that guaranteed poorer health all added up to a
population less able to fight off disease.

A *Nature Communications* paper from 2013 suggests that
when farming arrived in Western Europe 7,500 years ago, a
population 'boom-and-bust' pattern emerged. The number
of people grew rapidly and then crashed, then grew again.
No one knows exactly why the crashes occurred, but Steve
Jones has hypothesised that perhaps they were the result of
the first disease epidemics to hit humankind. It's an intriguing
possibility, but one for which there is no evidence to date.
What we do know is that the Neolithic Revolution, somewhat
surprisingly to me at least, brought about an abrupt decrease
in the standard of living. People suddenly had to spend every
waking moment toiling their lands to feed themselves and,
even then, they were likely to be vitamin-and-iron-deficient,
with rotting teeth and dodgy backs. The author of *Guns,
Germs, and Steel*, Jared Diamond, refers to agriculture as 'the
worst mistake in the history of the human race' and the origin
of 'the gross social and sexual inequality, the disease and
despotism, that curse our existence'. It's a somewhat cynical
view if you ask me, but hard to disagree with.

You can imagine Chapter 2's tales of migration and
diversification as the sowing of the seeds for an epidemic.
When the Neolithic Revolution arrived, those seeds started
to grow. Over thousands of years, TB would go on to
gradually wind its way through human societies. Towns
morphed into cities and populations expanded. TB clung on
tighter and tighter as the Roman Empire provided it with the
structure it needed to flourish in Europe. TB remained our
constant companion throughout the Middle Ages, growing
in strength from century to century. Human remains from
this period often bear the scars of TB, among other things.

Digitised Diseases is a fascinating project bringing human
bones from archaeological and historical medical collections
out of the specimen drawers and onto our computer screens.
Thousands of samples have been photographed and scanned
using 3D laser scanning, CT scanning and radiography, after
which video-game designers were unleashed to populate the

resource with digitised versions of the bones. The end results are 3D models that anyone with a computer (or smartphone) can use to learn more about the physical changes occurring as a result of numerous diseases. I've spent the morning playing with the bones, buoyed by morbid fascination and a mounting feeling of 'thank heavens for modern medicine'. One model is of an adult's first to tenth thoracic vertebrae – that's the top bit of the spine. The fourth to the sixth vertebrae have been completely destroyed by TB, causing the spinal column to bend over at a right angle as if someone has simply folded it in half. In real life, its owner would have had a severe hump on their back that may well have looked a lot like the triangular dorsal fin of a shark. Some vertebrae are badly damaged or reduced to small wedges of bones compressed between the others, leading to scoliosis – a snake-like sideways curvature of the spine. Several rib heads have fused to the vertebrae, and there's severe pitting and bone destruction over pretty much everything. I try to picture the actual human being who once housed this spine, but I can't do it. The damage to the bones is so far beyond anything I can really comprehend.

I called Don Walker, human osteologist at the Museum of London Archaeology, to ask about what TB infection does to the human skeleton. He explained that the skeleton reacts to infection by either losing or creating new bone. With Pott disease, the bone is literally eaten away and you can sometimes see smooth lesions that look a bit like someone's got inside with an ice-cream scoop. If too much bone is destroyed, the spine collapses forward on itself. But that's the only definitive way to diagnose TB in remains. Other lesions in different bones – everything from the elbow to the ribs to the hip joint – can resemble these scooped-out holes or they can be seen as areas where new bone has been laid down. But without the Pott disease of the spine, it's difficult to say for sure that other bone damage is the result of TB and not another infection. It amazes me that people survived, for a while at least, with such horrendous manifestations of their disease. Crumbling spines, destroyed hip joints. It's almost unbelievable that someone could be this ill and still alive. 'Evolution has made us pretty tough,' Don told me somewhat sadly. Referring to ancient remains, he continued: 'The only things that we see are the results of people suffering these diseases over a long period of time. Chronic diseases. These are the stronger people who've managed to survive the initial infection. The people who don't have any bone changes: while you might say, OK these people look healthy, these were the people who perhaps died after a week or two before there was time for the bone changes to take place.' It's perhaps this chronic nature of TB that left such a mark on human history. It wasn't just killing people, it was making a lot of people very ill for a very long time.

Many of the bones from the Digitised Diseases project date from the Middle Ages, when TB was one of many awful ways to die. There's a passage in one of John Bunyan's books, *The Life and Death of Mr Badman*, which reads:

He was dropsical, he was consumptive, he was surfeited, was gouty, and, as some say, he had a tang of the pox in his bowels. Yet the captain of all these men of death that came against him to take him away, was the consumption, for it was that that brought him down to the grave.

Captain Consumption aka TB wasn't the only condition that ailed poor Mr Badman, but it was the final nail in his coffin. I was reading John Graunt's observations on the bills of mortality for London from the late seventeenth century, when Bunyan wrote his book. Alongside the rarer 'Canker, Sore mouth and Thrush' and 'Found dead in the streets', there were entries for 'Grief', 'Frightened' and 'Stopping of the Stomach'. But the most common killer was, unsurprisingly, 'Consumption, and Cough' (TB), followed by 'Chrisomes, and Infants' (early childhood deaths grouped together rather than separated by disease) and 'Ague and fever'. In this pre-antibiotic era, around half of TB sufferers would be dead within a year, but of those who survived, half again would suffer from a chronic, lingering form of the disease. These would be the people who'd leave behind the scars of TB on their bones and inspire the eighteenth and nineteenth centuries' romanticisation of the disease.

In 2015, a study led by Thierry Wirth described how the Beijing strains of *M. tuberculosis* evolved into a pathogenic monster best known for spreading drug-resistant TB. Wirth and colleagues collected nearly 5,000 isolates from 99 countries, and sequenced the whole genomes of 110 of them. From this, they were able to reconstruct the Beijing Lineage's family tree, revealing where it underwent big growths in population size. These explosions of diversity are an indication of points in time when the disease was most successful. Living up to its name, the Beijing Lineage emerged near north-eastern China around 6,600 years ago, right when rice farming was getting under way in China's Upper Yangtze River valley. What's interesting, though, is that the Beijing strains underwent a huge burst of diversification at the time of the Industrial Revolution. Based on the graph in Wirth's paper, the estimated population size grew by around a factor of 10, indicating that it had found its niche among the human population and was taking full advantage of its comfortable surroundings. Comfort for *M. tuberculosis*, however, is indirectly proportional to comfort for its host. This brings me to what is simultaneously one of my favourite eras of history and somewhere I'd never want to visit: nineteenth-century London.

From a distance, the streets were paved with, if not gold, coal. An endless stream of migrants flocked to the city to find their fortune, or at least a job in one of the new factories. What began in Britain soon spread to other parts of the world, and cities saw an unprecedented growth in size. The population of London stood at 1 million at the beginning of the nineteenth century, rising to 6.7 million a century later. The city's infrastructure could barely keep up. Living conditions in the poorest districts were something straight out of a Dickens novel. Tenement slums with

narrow streets, teeming with crime and grubby-faced
children. Flapping washing strung between windows that
looked in on tiny rooms housing entire families. A heavy
smog of coal smoke and the stench of the sewage-polluted
Thames filling the air. Windows remained shut, providing
unnecessary protection against the (wrongly accused)
disease-causing miasma – air thought to be tainted by
exposure to rotting corpses, sewage or sick individuals.
Inequality between the rich and the poor grew; health
among those at the bottom of the pile deteriorated. From
what I've heard, it wasn't the greatest time to be a human.
'If you look at Roman bones or medieval bones, they look
pretty normal,' Don Walker told me during our conversation.
'But once you get up to the eighteenth or nineteenth
century, when you had the Industrial Revolution and
massive industry and more chemicals in the air and a lot
more overcrowding, you can see in the bones just how
unhealthy they were.'

Infectious disease spread through the population easily
and unforgivingly. The difference in death rates between the
rich and poor was stark. Perhaps one of the better-known
(albeit fictional) examples is Tiny Tim Cratchit from
Dickens's *A Christmas Carol*. Because of Ebenezer Scrooge's
stinginess, his long-suffering employee Bob Cratchit lacks
the means to buy enough food and medicine for his sickly
son. In one of his run-ins with the ghosts of Christmas,
Scrooge learns that Tiny Tim will soon die. But once Scrooge
accepts the error of his ways and starts paying his workers
decent wages, Tim is miraculously cured. Over the years,
many an internet sleuth, historian or doctor has attempted
to diagnose the character's condition based on his symptoms.
What disease could have crippled poor Tim and led to his
death in one future, while being completely reversible upon
Scrooge's conversion to a more charitable nature? An
amusing paper by Charles Callahan from 1997 describes the
imagined unearthing of one Timothy Cratchit's grave as a
result of excavation work on St Andrew's Church in London.
Callahan reported how workers had unearthed a grave stone

reading: 'In Memory. Timothy Cratchit. 1839–1884.'
Underneath, they found the skeleton of man 'approximately
40 years of age wearing a frame of metal and leather on his
legs and lower back.'

Could it have been that Timothy was suffering from Pott
disease? Callahan certainly thinks so. It fits with many of
Tiny Tim's symptoms. The short stature, pain and weakness,
difficulty in walking. A more recent review article by Russell
Chesney reaches a similar conclusion following an
exploration of the historical environment in which Tiny
Tim would have lived. The family hailed from the Camden
Town area of London and, based on the meagre '15 bob' per
week earned by Bob Cratchit, would have had a very limited
diet. In his review article, Chesney explains that this would
have purchased just four 1lb (450g) loaves of bread. A poor
diet combined with the perpetual smog blocking out
sunlight would have meant that many children of the time
were vitamin D deficient. This could have led to Tim
developing rickets, like 60 per cent of the children in
London, perhaps explaining the leg braces and crutch. Not
only this, but at the time, around 50 per cent of children
would also have shown signs of TB. It's a disease that presents
very differently in adults and children. So while Tiny Tim
never coughs or displays all the symptoms of pulmonary TB,
he may well have suffered from Pott disease.

In support of this theory, both rickets and TB could be
vastly improved by a better diet. More money would have
meant that Tim's meals could have been supplemented by
fish and dairy products. All of this would have upped his
vitamin D levels, curing his rickets and potentially sending
his TB into remission. Perhaps Scrooge's new role as Tim's
second father also went as far as to arrange for the boy to
travel to the countryside, where fresh air, sunlight and rest
could again have improved his symptoms. In the end, it
doesn't really matter what afflicted Tim, as the point of the
story is the transformative power of love, not cod liver oil.
If only real life was as heart-warming. Dickens's purported
inspiration for Tiny Tim was his nephew Harry Burnett,

who was to die of TB at nine years old. TB also claimed thousands of other nineteenth-century children, particularly those born to the poorest families. The link between poverty and the disease is illustrated by some alarming 1890 statistics from Hamburg, Germany. While 6.6 in 1,000 people earning below 912 marks died from TB, only 1.9 in 1,000 of the 10,000-marks-and-above earners succumbed. Nearly everyone in the nineteenth century would have been infected with *M. tuberculosis* by the time they reached adulthood, but whether or not someone exhibited symptoms or survived the active form of the disease could be strongly impacted by, in particular, their diet.

The sheer scale of the TB epidemic during the nineteenth century made it a staple part of culture. Only, the era's romanticisation of TB doesn't paint a very accurate picture of the disease. I've read again and again that TB was often credited with imbuing a person with a passion and creativity unachievable by those who'd yet to join the consumption club. In many examples of literature of the time, it is treated as a disease of moral superiority, capable of intensifying the experience of life. As the body wasted, consumed from within, artistic talent could feverously bloom and soar to new heights. With the roll call of famous TB sufferers including John Keats, most of the Brontës, Paul Gauguin, Frédéric Chopin, Robert Louis Stevenson and Franz Kafka, it's easy to see why this link between creativity and consumption was proposed. French writer Alexandre Dumas is quoted as saying: 'It was the fashion to suffer from the lungs; everybody was consumptive, poets especially; it was good form to spit blood after any emotion that was at all sensational, and to die before reaching the age of 30.' A mass killer of poets, then, but what a euphoric, passionate death! What I find hard to fathom is whether the people living with TB truly believed this. Maybe we're just seeing the nineteenth-century equivalent of Facebook – a lens through which life looks brighter and happier. Perhaps we've lost the edge of sarcasm to the poet Lord Byron's voice when he said, 'I should like, I think, to die of a consumption,' and when questioned further, 'because then the women would all say, "See that poor Byron – how interesting he looks in dying."'

The truly horrific aspects of the disease were often glossed over in the artistic works of the day. Operatic heroines such as *La Bohème*'s tragic Mimi or *La Traviata*'s slightly feistier Violetta sing passionately with their dying breaths; a final burst of stamina before expiring. The film *Moulin Rouge!* was based on the latter, and Satine suffers the same fate as Violetta, although with even more drama (she's literally on stage at the time). It's as if these characters are made into something bigger than themselves by their deaths from the disease. How different these productions would have been if they'd stuck to a realistic course of TB infection. Those final arias would have involved a lot of coughing and suffocating respiratory failure. But it wouldn't have fitted with the tragic but beautiful TB victim archetype of the day. When Edgar Allan Poe described his wife, Virginia, as being 'delicately, morbidly angelic', we can picture her, porcelain-skinned and melancholy, with cheeks rosy from fever and a brightness to her eyes. And when Poe remarked, 'Suddenly she stopped, clutched her throat and a wave of crimson blood ran down her breast ... It rendered her even more ethereal', it's easy to put his flowery descriptions of the disease down to his writer's nature. Although when you stop to think that this disease had already claimed the lives of his mother, brother and foster mother, his prettification of TB is suddenly a whole lot sadder.

Perhaps, like New England's vampires, the nineteenth-century's obsession with romanticising consumption as a fashionable way to die was, in fact, the easiest way to deal with death on such an extraordinary scale ... along the lines of today's war against cancer, with words like 'brave battle' and 'continues her fight' used to restore some of the human power to what must be a terrifyingly powerless situation. It's scary to consider that, by the turn of the nineteenth century, TB had claimed the lives of one in seven people who'd ever lived. Most families had been touched by TB, yet the scientific

understanding of the disease still had some way to go. It wasn't
until 1882, in fact, that German scientist Robert Koch
presented his discovery of the TB bacillus. 'If the importance
of a disease for mankind is measured by the number of fatalities
it causes,' he began, 'then tuberculosis must be considered
much more important than those most feared infectious
diseases, plague, cholera and the like. One in seven of all
human beings die from tuberculosis. If one only considers the
productive middle-age groups, tuberculosis carries away
one-third, and often more.' Koch's discovery marked a new
era of microbiology and scientific research. The old-fashioned
superstitions surrounding TB would gradually fade away, to be
replaced with what I think is an equally terrifying scenario – a
curable, well-researched disease that still manages to kill more
than a million people a year. Incidentally, if you're interested
in the history of TB science rather than the science of TB
history, then I would recommend Helen Bynum's *Spitting
Blood*. The early science that set the foundations for our
understanding of TB is beyond the scope of this book, but is a
fascinating area.

Come the end of the nineteenth century, the positive
impacts of the Industrial Revolution and all the other changes
of the day had started to take effect. After the 1880s, new
food regulations, increased pay, better diets and general
hygiene and improved living conditions had a knock-on
impact on the health of the population. By the turn of the
twentieth century, TB was on a rapid downward trajectory –
50 years before the introduction of the antibiotics that are
often mis-credited with the end of the European and
American epidemics. Soon after the mid-twentieth century,
hopes were high that TB's story would shortly be one of
historical curiosity. TB control programmes were disbanded
and science moved on. It was a complacency that would later
come back to bite us all. Because, like vampires, TB has a way
of reinventing itself for new eras and new populations. Some
of the story remains the same: urbanisation, poor living
conditions, inadequate diet, mass migration, natural disasters.
Other parts are new: overpriced drugs and diagnostics, the

rise of HIV, new wars and displaced peoples, drug resistance, crowded prisons, stigma. TB is no more a disease of the past than vampires are a forgotten piece of historical folklore. 'It never ceases to amaze me that it's a story format that keeps resurrecting itself,' Nicholas Bellantoni told me. I know he was talking about vampires, but he could just as easily have meant TB.

CHAPTER FOUR

All That Glitters

'You cannot destroy it because it is the lungs of the world,' says shaman and spokesperson Davi Kopenawa. He is talking about the Amazon rainforest, a vast ecosystem that produces about 20 per cent of the planet's breathable oxygen. Home to hundreds of mammalian, tens of thousands of plant and millions of insect species, Davi Kopenawa's place of birth is gradually being lost to cattle farming, mining, logging and subsistence agriculture. As the rainforest falls, so do the indigenous hunter-gatherers who have prospered among the lush vegetation for thousands of years. Davi's quote seems all the more poignant given that respiratory diseases such as tuberculosis are among the many threats to these indigenous people.

Davi grew up among the Yanomami – the largest isolated tribe in the Amazon, numbering some 32,000 people and occupying 9.6 million hectares (37,000 square miles) of land. The Yanomami had no sustained contact with the outside world right up until the 1950s, when Catholic and Protestant missionaries brought them bibles and a number of diseases to which they had no natural immunity. These good men and women of god inadvertently killed the very people they were trying to save, with epidemics of flu, measles and whooping cough. Davi's own parents were among those who died. Thirty years later, colonists discovered gold in the Yanomami territory and 40,000 miners descended upon the rainforest. They brought with them illegal landing strips, mining techniques that poisoned the rivers and fish, and genocide. Oh, and yet more diseases that spread through the Yanomami like a forest fire. Over just a few years, an estimated 20 per cent of the Yanomami died from respiratory infections, malnutrition and malaria introduced by the gold miners. Takes some of the shine off that gold jewellery when

you realise it costs far more than the current market value of
£32.83 per gram, doesn't it?

A charitable organisation called Survival is campaigning
for the rights of tribal people around the world, including
those living in the rainforests of Brazil and Paraguay. They
can tell dozens of stories of indigenous Indians who have died
from TB – people forced to make contact with outsiders after
bulldozers destroyed their homes. It's all disturbingly similar
to the fate of indigenous Americans wiped out following the
arrival of Christopher Columbus. This time round, though,
we get to follow this shameful decimation of the indigenous
people through the lens of modern science and the media.
And unlike the Native Americans, the Yanomami truly were
immunologically naive towards TB until outsiders arrived,
with devastating consequences. Between 1990 and 1996, a
study conducted at the request of the Brazilian Ministry of
Health and Brazil's National Indian Foundation (FUNAI)
tracked the TB epidemic among the Yanomami Indians.
Following a five-day trip up the Rio Negro and two days
trekking by canoe, Alexandra Sousa and her colleagues
reached the remote villages nestled among the purple-fruited
Euterpe precatoria trees. The purpose of their study was to
assess the impact of TB on the indigenous population in order
to inform control measures. What they found was evidence
for TB's huge selective pressure on the human immune
response to infection.

Sousa's team examined 625 people across five villages and
discovered TB unlike anything seen previously. For starters,
the Yanomami had an extraordinarily high rate of active
tuberculosis – 6.4 per cent. Along with the 1.28 per cent
mortality rate, this put the Yanomami epidemic on another
level to anywhere else in the world. In comparison, the New
York drug-resistant TB epidemic of the early 1990s was 100
times less widespread. On top of the sheer numbers of people
suffering from the disease, the Yanomami also seemed prone
to serious complications not seen in Europeans since medieval
times. Scrofula – or King's Evil, to go by the old-fashioned
name – is a TB infection of the glands, leading to large

swellings of the neck that can eventually rupture. Don't Google it, seriously. In the European Middle Ages, it was believed that the disease could be cured by the royal touch. Up until the eighteenth century, hordes of English or French sufferers would queue up to be touched by their monarch, with King Henri IV of France purportedly touching up to 1,500 victims at one time. Today, however, neither Queen Elizabeth II nor François Hollande performs this strange rite. Scrofula is extremely rare among HIV-negative adult TB patients, so I suppose there's little call for it. This condition of the past, however, is very much present among the Yanomami, with 18 per cent of TB sufferers showing signs of scrofula. Sousa's paper hints that something about the Yanomami's immune programming is different, meaning that they don't respond to a TB infection in the same way that you or I would.

In simplistic terms, you can think of the adaptive branch of the immune system as having two main modes. When it comes across something that shouldn't be there, it decides which mode to focus on based on how best to destroy the threat. These modes are the antibody response and the cell-mediated response – or as I like to think of them, Paint Ball versus Hide-and-Seek. First, we have the antibody response (PaintBall), which tags threats and marks them for clearance from the body. It's a weaponised branch of the immune system that sends out its drones (antibodies, antimicrobial peptides, complement system) to do the dirty work, while the cells of the immune system sit back and orchestrate warfare from a safe distance. The cell-mediated response (Hide-and-Seek), in comparison, uses a door-to-door approach to find host cells that have been infiltrated by the bad guys. The immune-cell sentinels peer through windows or look for empty pizza boxes on the lawn and then, should they find something dodgy, personally demolish the entire house, killing everything inside. In reality, of course, it's not as simple as one mode or the other. You need both to fight off most infections: the antibody response to combat any pathogens roaming free, the cell-mediated to find and destroy those lurking inside host cells. A whole lot of natural selection and evolution can teach the immune system how much of each mode to switch on in response to a given infectious agent. Simply put, those people whose immune systems randomly get it right get to live. They pass on the correct response to their descendants while everyone else dies, their unlucky immune programming dying with them.

When it comes to TB, the Yanomami make huge levels of antibodies and not much else. The strong cell-mediated response vital in the fight against TB is sluggish or non-existent. Sousa and colleagues hypothesised that, in populations exposed to TB over hundreds or thousands of years, selective pressure has shaped the immune system in ways that the Yanomami have not experienced due to their isolation from the disease. Infectious pathogens, after all, have been one of the major selective forces during human history. Traces of the effect of pathogen-mediated natural selection can be found in the genes of the human immune system, with different populations bearing the marks of the diverse pathogenic environments that they've been exposed to. In the previous chapter, I talked about how TB ripped through the European population during the seventeenth to nineteenth centuries, causing between 20 and 30 per cent of all deaths. Before then, the disease had been hanging around in humans for millennia. Surely something so prevalent and deadly would have exerted a strong selective pressure on genes capable of affording their humans some level of protection?

The fact that the European TB epidemics were already declining a century before anyone invented effective drugs certainly suggests that the population got better at dealing with TB, although social factors clearly played a huge role here. Could it be that part of the decline was due to selection for those genes responsible for triggering a strong cell-mediated response to the infection? It's not an easily testable hypothesis. Marc Lipsitch gave it a try in 2002, but concluded that natural selection by TB deaths couldn't account for the reduction in TB disease following the European epidemic.

Three hundred years wouldn't have been long enough to measurably increase the frequency of protective gene variants. All the same, I find it intriguing to imagine that I am the *After* and the Yanomami the *Before* in terms of our ability to fight off TB. With my European and North American ancestry, I would likely mount a strong cell-mediated response, control the infection to some degree and probably (hopefully) not develop scrofula. Again, it's not a hypothesis that I have any intention of testing.

History tells us that uncontrolled TB epidemics tend to last around 300 years, with the highest mortality occurring at the 50-year mark. The Yanomami are currently 35 years into their epidemic. Given enough time, perhaps natural selection would eliminate a high proportion of the susceptible individuals and enrich the gene pool with genes capable of protecting their humans from TB at its worst. Thankfully, today's anti-tuberculosis therapies mean that the Yanomami won't have to wait for evolution to do its thing at the expense of so many lives. It's just a shame that modern medicine needs to intervene to fix a problem that the outside world caused in the first place. I also need to mention that TB control attempts among indigenous populations are fraught with difficulties. In the past, these groups have displayed high rates of treatment non-completion and higher levels of fatalities compared to the non-indigenous population. Part of the problem is that the health service isn't stretchy enough to reach their secluded villages, especially not when it comes to a disease like TB that requires several months of closely monitored treatment. Recent efforts have included the decentralisation of health services to bring medical care to the communities that need it and the introduction of DOTS (Directly Observed Therapy, Short-course) in the late 1990s. In one study, the incidence of TB among indigenous Brazilians in Dourados had halved since the year 2000, and the non-completion of treatment rate had been reduced by 90 per cent. There is still a long way to go, however, and further reductions will likely require more active case-finding to identify patients early before transmission occurs.

As more studies come out looking at TB in indigenous populations, one thing that is becoming more obvious is that trying to blame the differences in TB progression and incidence among these people solely on their genetics would be extremely naive. These groups of people live in extreme poverty, with reported poor health, including high rates of parasites, malnutrition, anaemia and other infections. These medical problems are compounded by food insecurity, unemployment, low or no income and shared living spaces with poor ventilation. These are the same factors at play as during the nineteenth-century TB epidemics in Europe and North America. Like many places in the world, ethnicity isn't simply coded in your genes but is twisted up with complicated social constructions and socio-economic inequalities. It's interesting from a scientific perspective to speculate on how the immune naivety of people such as the Yanomami contributes to their TB epidemic. But speculation is as far as it goes, and real-life TB control has to take into account all the real-life stumbling blocks to providing healthcare to these populations: culture and language barriers, differences in their conceptions of how disease works, poverty and poor living conditions that encourage transmission. TB is almost without exception so much more than a man-versus-microbe medical problem.

Where studies of genetic susceptibility to TB can help us, though, is in understanding how to improve on the natural immune response in the development of new vaccines or treatments. You have to remember that, for most people, the immune response does a good job of keeping TB in check. Of those exposed to the bacterium, around 10 per cent will completely fight off the infection. Of those who can't, 95 per cent will do a good enough job to force the bacteria into a latent infection. Of these, only 10 per cent will progress to active TB during their entire lifetime. Based on these numbers, nearly 90 per cent of those exposed to TB will never become ill. So what's so special about these people that protects them from the disease? Some of the earliest work on the genetics of TB susceptibility was performed in the 1940s by a German called Franz Kallmann. Great scientist; bit of an arsehole. At one point, he made the lovely suggestion of examining relatives of schizophrenia patients so that 'non-affected carriers' might be sterilised and the disease-causing gene removed from the gene pool. A year later, his Jewish heritage forced him to flee Germany for the United States, leaving behind his ex-boss, Ernst Rüdin, to pioneer the Nazi idea of racial hygiene and eugenics. Anyway, Kallmann was at the forefront of the use of twin studies in looking for a DNA-based explanation for disease.

TB, interestingly enough, was originally thought to be a genetic condition. So strong was this prevailing belief that back in 1868, when Jean-Antoine Villemin had demonstrated that he could transmit TB from humans or cattle to rabbits, no one had believed him that TB was a contagious entity. TB, you see, ran in families. Take the Brontës, for example. Six people – an entire generation – suffered from TB, and most, if not all, died. Sounds like an inherited disease, right? It amuses me that our understanding of TB has almost come back full circle, as today, while we know that the disease isn't genetic, it's accepted that our response to it certainly is. Some of the first evidence for this came from Kallmann's studies. In 1943, he published a paper with the distinctly uncontroversial David Reisner looking at the incidence of TB among twins. They examined 308 twin pairs and found that monozygotic twins were more likely to share TB than dizygotic twins. Basically, if you're a monozygotic twin and your counterpart develops active TB, you're far more likely to get the disease yourself than if you'd shared a womb but not an egg. The closer people are genetically, the more likely they are to have comparable susceptibility to TB. But what genes (or hundreds of genes) are responsible for this genetic susceptibility and how can we find them?

Back when I was a TB scientist, I was a member of the Acid Fast Club. My partner finds the name of this professional society hilarious, but it's not as exotic as it sounds. There are no class A drugs; no house music. Just a whole load of leather satchels and the occasional beard. The Acid Fast Club (so named because of a staining method used to identify mycobacteria) holds two meetings a year, in which scientists present their work to an audience of mainly UK researchers. There are biscuits in the interval and a trip to the pub at the end of the day. It's all very nice and sciencey. Anyway, the reason I'm bringing this up is that for several years the meetings were besieged by research attempting to identify human TB susceptibility and resistance genes. My heart used to sink every time I saw one of these talks on the programme. Chances were, they'd start out really well, daring to raise my little hopes. Look, a region on chromosome 7 that appears to be linked to a person's ability to prevent TB! Ohhh, and it's a region that contains a number of genes involved in the immune system. I'd start to see the headlines: *Scientists discover the gene for TB. Boffins cure TB.* That sort of sensationalist nonsense that, despite my better judgement, fills me with pride over my fellow researchers' hard work. Then one of the final slides would only go and announce that the statistics weren't up to scratch. While the results were *interesting*, they weren't statistically significant. This is the science equivalent of almost-but-not-quite getting three numbers on the lottery. So close, but no prize.

Most of these talks were about genome-wide association studies (GWAS). The basic idea is that you take discrete groups of people – such as those with active TB versus those who maintain the infection in its latent form – and you sequence their genomes. Then you mine through all that data to find something in their DNA that might explain the differences in infection outcome. A genetic marker that's associated with the trait you're interested in. Ideally, it would be one single gene that comes in two flavours: *TB* or *Not TB*. Once you've found the gene, you can get to work finding out what its biological function is. Sounds easy, right? It would be, if all diseases were as simple as, for example, Huntington's disease, where a single faulty gene really is all it takes. But susceptibility to complex diseases like TB is a messy combination of genetic and environmental factors, with hundreds of genes working together to increase susceptibility by maybe as little as 10 per cent. There could be a dozen or a hundred or a thousand different evolutionary solutions to the same problem (how to survive TB), all of them involving different combinations of many, many genes. From a GWAS point of view, the more complicated a disease, the more people you need to include in your experiments to stand any chance of noticing a correlation. And with TB, there's an additional problem, in that the lines between the different groups of patients are blurred and hard to define. So TB GWAS experiments have proved notoriously difficult to reproduce and have generally failed to confirm pre-GWAS candidate genes thought to have a role in TB susceptibility. I can vividly remember sitting through what felt like hundreds of talks (it was probably more like five over as many meetings), most of which ended up explaining how maybe their population of patients was different to previous studies, or maybe their definition of latent TB was different, or maybe blah, blah, blah. But the gold standard in any genetic association study is replication. When you're looking at such small effects, you (or someone else) need to be able to repeat the results. Otherwise you can't prove that the data are anything more than expensive noise – the scientific equivalent of hearing voices in the static.

That's not to say there haven't been some success stories. Sergey Nejentsev went for the Go Big or Go Home option in 2015, with the largest TB GWAS to date. This study of 15,087 patients in total identified an association between pulmonary TB and a gene called ASAP1 involved in the migration of immune cells. In addition to Nejentsev's work, there are other regions of the genome where the evidence is stacking up in favour of their involvement in TB susceptibility or resistance, often in specific populations. But the relatively small contribution of individual genes means that it takes tens of thousands of patients to pick out a clear effect. I talked to Adrian Hill of Oxford's Jenner Institute, author of two of just a handful of convincing TB GWAS papers. He was confident that we will soon start to see more definitive answers thanks to global collaborative efforts that allow researchers access to greater numbers of patient genomes. 'We're probably not going to end up with a kit that I can test you with and tell you how susceptible to TB you are,' he told me. 'What we're trying to do is to understand better the vulnerable points of TB where amplifying host immunity or tilting it in one direction to a different sort of immunity would make a big difference to TB control.' Here, he's referring to how, of 100 people exposed to TB, 90 per cent will never become ill. 'What's different about the 10 per cent compared to the 90 per cent exposed to the same dose of the same strain of TB?' he asked. 'Most of us have the right genes that let us not end up with tuberculosis. There isn't a huge difference in genetics between different people, so it might be quite a small modulation that is all that's needed to turn the 10 per cent into the 90 per cent.' It's this idea of giving the immune system a shove in the right direction that intrigues me the most. But it all relies on working out what parts of the immune system are the important parts when it comes to TB.

The progression of a TB infection is an intricate ballet of migrating immune cells, dividing bacteria and host-mediated inflammation, the exact roles of which are not yet fully defined by science. No wonder, when you consider that this interplay between man and microbe has been plotted out over centuries of evolution. Along the way, the relationship has become entangled with a wealth of social and environmental factors, and in just the last few decades has been knocked off balance by one of the newest threats to TB control – HIV infection. Remember how I was saying only 10 per cent of those with latent TB will progress to active disease during their lifetime? Well, that's pushed up to 10 per cent *per year* when HIV infection enters into the equation. It's like, we've had all this time for our immune systems to find clever ways to deal with TB and, for most of us, it's really successful in ensuring we remain alive. Only, then along comes HIV and makes all of that kind of pointless. I'll come back to how, specifically, HIV messes with our immune system later. For now, though, I want to look at an example of a way in which HIV is unbalancing the co-evolved equilibrium between

M. tuberculosis and us humans. So, let's trek nearly 5,000km (3,000 mi) from the mist-shrouded trees of the Yanomami's rainforest to the foggy skyline of San Francisco. Here, the landscape of TB infections changes dramatically. But, like the Amazon, it was gold that brought the disease to the city, carried in on overcrowded clippers as thousands upon thousands of miners arrived during the Californian gold rush of 1849. 'There's gold in them thar hills!' and, also, hastily erected settlements rife with disease. By 1855, something like 300,000 fortune seekers had arrived in California, populating the rapidly expanding San Francisco with people from all around the world and the various diseases that came along for the ride. Sepia photographs and illustrations from the day have a nautical Wild West vibe to them. Ships making a forest of masts in the bay, square buildings lined up on the dirt, clusters of tents and wood shanties. Much like London during the Industrial Revolution, thousands came to the city to find their fortune. Instead, many came face to face with TB as the disease flourished thanks to overcrowding and poor health among the population.

Today, San Francisco remains a cosmopolitan centre of immigration, both human and microbial. It's a mixing pot for different ethnic backgrounds and for strains of *M. tuberculosis* originating from diverse locations. The gold rush was responsible for some of the first Chinese immigrants finding their way to San Francisco, who braved several perilous months voyaging across the Pacific. A plane can make the same journey in around 12 hours these days, and San Francisco's Chinatown has grown into the largest outside Asia. As globalisation gradually homogenises global culture, with Starbucks seemingly leading the march towards identical cups of overpriced coffee across world, I got to wondering if global traffic is also beginning to smooth out the diversity within the *M. tuberculosis* lineages. In 2005, Peter Small of Stanford University published the results of a study looking at 875 strains of *M. tuberculosis* isolated in San Francisco over an 11-year period. He found six different lineages, all making themselves feel at home in one of the most diverse cities in

America. People born in the US, China, the Philippines, Central America and Vietnam make up the five main populations of TB patients in San Francisco. In Small's study, each group had its own personal profile of *M. tuberculosis* lineages. US- and Central America-born patients were most commonly infected with the Euro-American Lineage, China-born with East Asian and Philippines-born with Indo-Oceanic. These are the same lineages that predominate in the US, China and the Philippines. So San Francisco is a mirror for the distribution of *M. tuberculosis* around the world, reflecting a fondness of specific strains for specific human populations that has yet to be lost to globalisation.

What was really interesting about the study was that Small's team looked at clustered cases – outbreaks in which more than one person was infected with the same strain. This allowed them to identify cases resulting from recent transmission rather than reactivation of an infection acquired in a patient's country of origin. The association between a person's race and the local *M. tuberculosis* lineage still held. Thousands of miles from home, lineages of *M. tuberculosis* were still more likely to infect hosts originating from the same part of the world. Within each population, the 'home' strains seemed to spread more easily than 'foreign' strains. So, a China-born patient infected with the East Asian lineage was more likely to transmit their infection than if they were infected with the Euro-American or Indo-Oceanic Lineage. Patients born in the US, Central America, the Philippines or Vietnam infected with the East Asian Lineage were more likely to keep their infection to themselves. Considering the fact that the East Asian Lineage has been introduced and reintroduced to San Francisco since the first Chinese immigrants arrived looking for gold, it's had plenty of time to spread. The fact that it has stuck to its preferred flavour of host is telling. It implies that *M. tuberculosis* has adapted to infect specific populations of humans and struggles if it comes across someone whose immune system doesn't quite behave as it's used to. You can reach a similar conclusion when you consider the West Africa 1 Lineage, which is very strongly geographically restricted to parts of West Africa. This lineage appears to prefer infecting those of the Ewe ethnic group and, although it surely must have been introduced into the Americas by the slave trade, it never took hold among Americans of European descent.

I like this idea that we humans have evolved ways of dealing with *M. tuberculosis* that can only be overcome by the bacterium following many hundreds of years of co-evolution. The type of immune response switched on by the human host, the length of the latent period, the ease with which the bacteria can divide in the lung – fine-tuned characteristics unique to each human–lineage pairing. But all those carefully balanced interactions fall away once HIV is added into the equation. A while back, Sébastien Gagneux and colleagues in Switzerland performed a nine-year study in HIV-negative and -positive patients, examining 518 people and their personal *M. tuberculosis* infections. In HIV-negative people, the patient's place of birth was a good indicator of which lineage they would be infected with. Transmission was most likely to occur when the patient's genetic heritage and local strain were matched up (same as in Small's study). However, the relationship broke down when HIV was included in the analysis. HIV-positive patients were not only more likely to be infected with the 'wrong' lineage, but this effect was a function of their CD4 T-cell count (indicating the degree of immune suppression). Turns out, there's nothing like immune impairment to break down barriers.

When it comes to HIV making TB worse than ever, there are few better examples than that of the South African gold mines. Today, South African miners have an incidence of TB that almost rivals that seen in Sousa's study of the Yanomami. It's no surprise that HIV infection plays a huge role in the problem, increasing the miners' risk of developing active TB by as much as five-fold. One study suggested that one-third of men become infected with HIV within 18 months of working in the mines. The single-sex hostels in which the men live – away from their partners and families – are hotbeds of alcoholism and prostitution. HIV rapidly spreads among both the miners and the sex workers and, later, when the miners return home, will spread to their partners. Swaziland and Lesotho are among the worst-hit countries in the world when it comes to HIV. They are also countries where a huge number of men migrated to South Africa for work and later brought the virus back home with them.

I was looking into the issue of TB among miners when I came across a 1945 poem by South African Zulu poet Benedict Wallet Vilakazi, translated into English by Florence Louie Friedman. 'Ezinkomponi' angrily talks of working in South Africa's gold mines and is uncomfortable reading. Part of it says: 'Roar! and roar! machines of the mines | Our hands are aching, always aching | Our swollen feet are aching too | I have no ointment that might heal them | White man's medicines cost money | Well I've served the rich white masters | But Oh, my soul is heavy in me!' And here lies the problem – segregation and, later, apartheid. South Africa's mining industry not only underpinned the modern economy, it also set the scene for much of the country's troubled history of racism. When gold was discovered in South Africa in the late 1800s, it led to the mass migration of predominantly poor black men from rural areas to work in the mines. Not only was there a need to ensure that the workers did not settle in 'whites-only' areas, but looking after the health of long-term workers would eat into the mines' profits. And workers would invariably become ill. By day, they toiled for long hours in the poorly ventilated, dusty and dangerous conditions below ground. By night, they returned to overcrowded quarters where disease rapidly spread among the weakened men. So the idea of circular or oscillating migration was born, in which men only worked for fixed periods of time before returning to their rural 'homelands'.

It was great for profits! The rural regions provided a constant stream of new men to replace those who could no longer work. Sure, many of the returned miners went on to succumb to TB and various other diseases picked up as a result of their work. But it didn't matter. The mining companies had no responsibility for ex-employees who were laid off when they became ill. Dispensable labour, sent home to die. The most horrifying part? This is still happening. Today, three medical conditions act together to create something far worse than the sum of its parts: TB, HIV and silicosis. This third condition is the result of inhaling tiny particles of silica dust that gradually result in inflammation and scarring of the lungs. It's irreversible and often fatal. The condition also increases the chance that a person will develop TB. One paper states that HIV-positive miners are five times more likely to develop active TB, while those with silicosis have a three-fold increased risk. Together, though, HIV and silicosis increase the chance of TB by 15-fold.

As far back as the turn of the twentieth century, the problem of miners' health was already well documented. In 1914, the US Surgeon General presented a report on his inspection of the gold and diamond mines in a region of South Africa and Rhodesia. His condemnations included long working hours, poor diet and cramped quarters in which up to 60 labourers slept in three-tiered bunk beds. The men were dying at an alarming rate from pneumonia, TB, meningitis and enteric fever. Epidemics were inevitable 'in places where large numbers of men are herded together without sufficient breathing space'. Simple solution: improve living and working conditions, improve diet, improve medical care. This didn't happen. Today, in 2016, this cycle of men migrating to the mines, contracting HIV/silicosis/TB, becoming too sick to work, being made redundant and heading home too poor to afford treatment continues. Aaron Motsoaledi, South Africa's health minister, has been quoted as saying:

> If TB and HIV are a snake in southern Africa, the head of the snake is here in South Africa. People come from all over the Southern African Development Community to work in our mines – and export TB and HIV along with their earnings. If we want to kill a snake we need to hit it on its head.

Leaders of 15 South African nations have pledged to confront TB in miners through investment in control programmes, and progress is now being made. In May 2016, South Africa opened the door to class-action suits seeking compensation for those who contracted silicosis and TB as a result of working in the mines. It's estimated that there are half a million men living with the two diseases at present, on top of all those who have already died. Some of the mining companies will appeal the court ruling, and it's likely that it will be some time before many of those affected begin to see some payouts, but it's a start. This is on top of increasing efforts from many mining companies to provide their existing workers with free diagnostics and healthcare, and intervention programmes to prevent the spread of diseases like TB. But with 9 out of 10 miners thought to already be infected with the disease, stopping the epidemic is going to be a long slog.

In 2014, the *New England Journal of Medicine* published the results of a clinical trial aimed at reducing TB in the gold mines. The idea was based on a 1960s study in Alaska demonstrating that preventatively treating all household members with isoniazid results in a 55 per cent decline in TB incidence over six years. Would a similar idea work in gold miners? The new trial sought to provide screening for the entire workforce and to treat those with active disease, while providing nine months of preventative isoniazid for those without active disease. It was a double-pronged approach. Quick diagnosis and treatment would reduce transmission; preventative isoniazid would reduce reactivation cases in those with latent disease. It didn't work. While those employees on isoniazid preventative treatment had a reduced incidence of active TB, the effect was lost the moment the treatment was discontinued. Either the miners were not taking the drug for long enough to cure a latent infection, or they were immediately being reinfected upon completing the course. Until the issue of overcrowded hostels is fully addressed, TB will continue to transmit very rapidly among this population. Those miners at high risk of active TB – those with HIV and/or silicosis – would likely need to take isoniazid continuously to remain protected.

What this study does demonstrate is that the solution is going to require more than medical interventions. It's also a worrying indication that HIV is completely changing the landscape of TB. In the last chapter, I mentioned a paper by Thierry Wirth tracking *M. tuberculosis* populations over time. Remember the big explosion in population diversity during the Industrial Revolution? A hundred years later, the global burden of *M. tuberculosis* appears to be going up again, coinciding with the rise of HIV.

Thanks to HIV, the long-term co-evolution of humankind and *M. tuberculosis* is entering a new phase, and I'm not sure what the future will bring. It's humbling to think that after millennia of co-adaptation between us humans and the TB bacillus, this delicate balance can be completely overturned by a virus that has been with us for only 100 years. It's one of the reasons why the race to discover new therapies and vaccines has never been more important. Thanks to HIV, latent TB is a ticking timebomb. However, at the moment we not only struggle to treat latent TB but also to diagnose it in the first place. Trying to prevent a disease that we don't fully understand is like trying to mine for gold with no map. If we don't know what constitutes a protective immune response against TB, then surely we risk swinging that metaphorical pickaxe at rocks and getting nowhere? Research such as GWAS, however, has the potential to inform us on where we should start looking. What we can be sure of is that as we move into a new phase of co-evolution between man and microbe in which HIV and drug resistance continue to make everything more complicated, we are running out of time to hit gold.

CHAPTER FIVE

Thanks for the Memories

Welcome to 1993. I've recently had my hair permed, become a vegetarian and decided that I want to become a research scientist when I grow up. Like Louis Pasteur, only less masculine and more not dead. Part of this decision has been triggered by the lovely weeping sore on my left bicep, which several weeks on is still gloriously green and crusty. All of us Year Nines, with our floor-length brown kilts and untucked shirts, have similar wounds. Mine is one of the biggest, much to my delight. I'm the queen of the ulcerating lesion and I don't care if you think it's disgusting. Twenty years down the road, the scars of 1993 still persist. While most are of the inwardly cringing variety (a perm, really?), the 8mm (⅓in) blemish on my arm is a physical reminder of a time when all 10- to 14-year-old UK school kids were vaccinated against tuberculosis.

Bacillus Calmette-Guérin, or BCG to its friends, was isolated by two very patient Frenchmen – Albert Calmette and Camille Guérin – over 13 years and 239 agar plates. I like to imagine this experiment as the bacterial equivalent of spending 13 years sitting on a sofa, becoming increasingly out of shape, then attempting to run a marathon. Only in BCG's case, the sofa was a delicious mixture of glycerine, bile and potato, and the marathon was the ability to cause TB. Plate 1: *M. bovis* (cow TB). Plate 239: a much-weakened shadow of its former glory. BCG is what's known as a live attenuated vaccine. Attenuated, like a T-rex without its teeth, or my hair after its run-in with the perming solution. While BCG is capable of inducing an immune response in humans along the same lines as a TB infection, it doesn't cause disease in healthy recipients. This is because somewhere between plates 1 and 239, the bacterium acquired a number of mutations in genes required for virulence. It didn't need them at the time, what with its home being an agar plate, so their functions were lost.

Modern-day vaccine regulators would have laughed BCG out of the room. Highly pathogenic parent strain? Unknown attenuating mechanism? Undefined mechanism of action? Morally questionable human experimentation? Yeah, get lost. No matter, though, as since its introduction, BCG has proven itself (thanks to a hefty dose of good luck rather than clever design) to be extremely safe. It has tallied up 4 billion doses over the years, and 120 million people, mainly children, continue to receive it every year, making BCG the world's most prolific vaccine. So you'd be forgiven for thinking that it works. It kind of doesn't. In clinical trials, the vaccine could prevent anywhere between 80 per cent and zero per cent of TB cases. Yes, you read that right. In some parts of the world, particularly those close to the equator, BCG does absolutely nothing to prevent people from developing TB. The 80 per cent figure came from a 1956 trial in the UK looking at vaccination of adolescents. So good news for me – or it would have been if there'd been much TB floating around in 1990s Bishop's Stortford (nope). But my counterparts in Chengalpattu, India, don't fare so well. Here, the vaccine provides no protection against TB at all.

Are human genetics at play, or is it down to something in the environment? Probably both. Exposure to environmental microorganisms, in particular, is one of the big suspects. One theory is that naturally occurring mycobacteria effectively vaccinate a person on the sly. This either primes the immune system to rapidly destroy the BCG strain before it can have an effect, or masks the effects of BCG by providing some protection against TB. Environmental mycobacterial species are more common in TB-endemic regions, while in semi-rural Bishop's Stortford there's not so much going on (in any sense). Basically, the underlying activation state of the immune system may well determine how effective a vaccine is. This could be the result of exposure to environmental microorganisms, a genetic quirk of how the human immune system is programmed, co-infection with other species, or something else entirely. The differences in immune system activation between populations or individual patients are something to bear in mind when it comes to making new vaccines, as clearly we want to protect everyone, not just those with cooperative immune systems.

This raises one somewhat obvious question: why is BCG still given so widely if it doesn't work? Ah, but I only said it *kind of* doesn't work. BCG is actually pretty good at protecting children, especially against more serious forms of TB that aren't content with remaining in the lungs but spread to other important parts of the body such as the brain. There's no denying that BCG saves lives, and it's therefore still given to infants in high-risk regions, such as sub-Saharan Africa and North London. UK teenagers like 1993-me are no longer inoculated, but at-risk infants are supposed to be. However, for the past few years there has been a global shortage of the BCG vaccine. Surprisingly, this hasn't received half the media and scientific coverage that I would have expected. In fact, most of the fuss kicked up seems to have come from the bladder cancer therapy field, where BCG is a key treatment.

The most worrying part of the shortage hit UNICEF, the main providers of BCG to high-burden countries. A few years ago, two of its four suppliers experienced technical difficulties that resulted in UNICEF being short by 8 million doses in 2013, 23 million in 2014 and 17 million in 2015. They did recently announce that the 2016–2018 period is fully covered, so we can hold off on the panic for a while. What all of this does highlight is that vital vaccines such as BCG need dedicated manufacturing investment to ensure that shortfalls like this do not happen again. But who is going to pay when the vaccine is sold for just cents per dose, making it a poor investment for pharmaceutical companies? Especially considering that two replacements for BCG are making their way through the development phase and should usurp the original BCG at some point in the future.

Funding is a problem that carries over into the development of new ways of preventing TB. Vaccine discovery is a high-risk endeavour with low potential returns. It takes 11 to

16 years for a candidate to make it from preclinical studies to phase III human trials, which alone cost an estimated $100 million. Phase II trials come in at a more modest $20 million. The WHO recently tallied up how much money is needed globally for TB research and development and came up with a figure of $2 billion. This would pay for drug discovery, diagnostics development and vaccine research. It's a hefty price tag, which is not being met by the current $0.6 billion available. On the vaccine front, the shortfall is around $250 million a year. In comparison, work towards an HIV vaccine received close to $1.2 billion during 2015. What the HIV policy makers seem to understand better than those responsible for funding TB is that a vaccine might be expensive, but these costs are far less than the global price of treating sick people (and that's just the economic cost, without any of the humanitarian considerations).

One of the challenges faced by the TB vaccine field is that it is built on a lot of unknowns. Why doesn't BCG work in some populations? What cells of the immune system are involved in successfully fighting off TB? What aspects of the TB bacillus best trigger the right response? How can we induce a long-lasting cell-mediated rather than antibody immune response through vaccination? How should we be testing vaccine candidates? All these unanswered questions recently proved too heavy a load for the TB vaccine field. It ended up snapping in two, and two camps emerged. Loyalists in the red corner and revolutionaries in the blue. I suspect that the revolutionaries would call their change in focus a much-needed paradigm shift, while the loyalists would probably accuse the revolution of being a knee-jerk response to a disappointing – but not fatal – setback.

I don't believe either is entirely in the wrong. For as long as I worked on TB there were mutterings over whether the vaccine field was heading in the right direction. I can remember watching the simmering tensions escalate from the relative safety of the TB drug-discovery field. Then, back in 2013, I attended a TB conference in Whistler, Canada, titled 'Understanding your Enemy'. One of the invited speakers was the UK's Helen McShane, who months previously had published the results of the first big TB vaccine trials since – well, since BCG. Lots of people had high hopes for her vaccine, MVA85A. In reality, though, it proved to be one of the more public 'failures' in recent TB research. I will just say that I personally don't like the word failure. It implies that the researchers involved did something wrong when, in fact, every scientific discovery is built on top of a thousand experiments that didn't work but that still contributed to of the field. However, the F-word was in abundance when the media reported on McShane's phase IIb trial. 'Experimental tuberculosis vaccine fails to protect infants', 'New TB vaccine fails trial', 'Key TB vaccine trial fails'. The fact is, though, that MVA85A didn't work, at least not well enough.

So I was really interested in hearing Helen McShane's lecture. My own opinion coloured by all those news reports, I was unsure how she was going to stretch out 'Nope, not happening' to fill an hour. To my surprise, though, the main gist of her talk was that we needed to learn from MVA85A and focus on the positives. At the time, I was unconvinced. Like many scientists in the room, I was concerned that the foundations upon which the entire field rested had been built in the wrong place, even if Helen McShane wasn't quite so pessimistic. It takes real belief in your work to stand up for a strategy that many people no longer agree with. I know it's not something I could have done, and it was in part this realisation that led to me leaving research less than a year later. With slightly more significance for the world at large, the vaccine field would also go on to change direction, despite Helen McShane's loyalty to the cause.

MVA85A was intended to be given as a booster to improve on the protection already provided by BCG. It comprises the Vaccinia Ankara virus engineered to express an *M. tuberculosis* protein referred to as Antigen 85A. You know how most of us have one feature that people use to describe us? *You know, Kathryn? That girl with the dodgy perm?* Antigen 85A is *M. tuberculosis*'s perm. It's known as an immunodominant antigen (an antigen is a short, distinctive protein sequence on the immune system's watch list). If you look at the immune response to a TB infection, a large proportion of it is directed at Antigen 85A. It strongly triggers the cell-mediated immune response that I've already talked about, switching on CD4+ T-cells and inducing lots of IFN-y. You don't need to understand exactly what this all means, only that inducing this kind of response was the aim of many potential TB vaccines. MVA85A did a good job of this in the UK trials but when it came to the big South African trial, the response was 10-fold lower. Would it have worked as a vaccine if it had induced a stronger response, or are antigen-specific T-cell responses a red herring when it comes to vaccinating against TB? I'm not sure anyone knows for sure, but this is where the split in the vaccine field occurred all the same. Much of the dissent was based around the fact that the already skinny TB vaccine pipeline was populated by lots of vaccine strategies that overlapped with MVA85A. And if MVA85A hadn't worked, then what did it mean for everything else?

So, it was proposed that everyone go back to the drawing board and explore some wider concepts and ideas. Instead of just focusing on generating a straight-down-the-middle CD4+ cell-mediated response, maybe we should consider switching on other cell types. Rarer cells that might not comprise the bulk of the immune response to TB but that could make all the difference to whether the infection is cleared or not. Later, the field would also start to look more closely at antibodies, which while not being sufficient for protection against TB, are still involved in fighting off the disease. All this is a good thing for a TB vaccine pipeline that rivalled UKIP when it came to its lack of diversity. What isn't so great is using one setback as a reason to hesitate over performing new clinical trials to the point that funding starts to dry up. Talking about the MVA85A trial, Helen McShane told me, 'It was extremely optimistic to think that this would get a single answer from the first candidate we tested. If one looks at the fields of malaria and HIV – also both difficult pathogens to get vaccines for – the fields are a graveyard littered with failed candidates. But actually, it's my view that the only way we will develop a better TB vaccine is by testing it in people.'

Much of the hesitation to fund such trials comes down to not knowing what a successful vaccine-mediated immune response against *M. tuberculosis* looks like. If we knew this, we could test if a vaccine was generating the right kind of response in a human without having to wait years to see if they developed TB or not. This is known as a correlate of immunity. It's like how pilots check all the sensors and systems on an airplane before they take off so that they don't have to crash before they discover something isn't working. For MVA85A, the proposed correlate of immunity was a strong CD4+ T-cell response with lots of IFN-y. Unfortunately, while MVA85A did a good job of generating this kind of response, it didn't translate into protection against TB. I've often heard it stated that we need to find a correlate of immunity first before we start testing vaccines in people. However, when I talked to Helen Fletcher – an immunologist involved in the MVA85A trials – she told me that there isn't any vaccine in the world that's been developed using an immune correlate. This, she explained, is something that usually comes from your first successful trial, not before. You then use your correlate to refine the vaccine and take it into other populations without the need for a massively expensive phase III trial every time round.

'The way vaccine development works is you iteratively test lots of candidates until you find one that works. The problem is we don't do this in TB as it's too expensive,' Helen Fletcher told me. 'I think we should just bite the bullet and we should just do it.' Only, $20 million is a lot to gamble, so a reliable way to narrow down vaccine candidates would be invaluable. Vaccine developers working on malaria or flu, for example, can use human challenge models to test their vaccines. According to the internet, you can earn the grand sum of £3,000 or thereabouts if you're willing to give up two weeks of your life and be infected with the flu (and take an untested medication too, of course). Flu rarely kills healthy adults, though, and there are decent, quick-acting drugs to combat malaria. TB? Yeah, you can't go around deliberately infecting people with *M. tuberculosis* – not if you want to keep your medical licence and freedom. There are now a couple of human challenge models in the works, using BCG or an attenuated strain of *M. tuberculosis,* both of which briefly survive in the human body but don't cause disease. Sarah Fortune of Harvard Medical School is currently working on a five-year project to programme an attenuated strain of *M. tuberculosis* with a kill switch. The idea is that it can only survive a certain number of cell divisions before dying, meaning it can safely be given to people without the risk of an unwanted infection taking hold.

Another option is to infect non-human primates, specifically macaques – something that JoAnne Flynn has pioneered at the University of Pittsburgh. This has the added benefit that you can put the animals in a CT scanner and determine quickly if a vaccine is having an effect by directly looking at areas of infection in the lung. I talked to JoAnne about her work – and I'll talk more about it in a subsequent chapter – but, when it comes to vaccines, she is hopeful that imaging is the way forward in testing if a candidate is going to work. In macaques, it's possible to visualise the effects of vaccination within months of TB challenge. Helen McShane's trial, in comparison, followed patients for over two years. Lots of people believe that this non-human primate model is the future of vaccine development, allowing scientists to narrow down candidates into those most likely to succeed in people. Sounds like a no-brainer, doesn't it? Only, the price tag is still around $5 million per candidate tested. 'I would actually argue that it's four times more valuable to test the vaccine in humans than it is in a non-human primate,' Helen Fletcher says somewhat controversially. 'And then you can at least take samples and learn about the disease, and raise awareness in your population.' It's worth noting that, while MVA85A didn't work as a vaccine, the efficacy trial has gone on to yield 10 additional papers in the years since. And what vaccine trials achieve is far more than science. They help to set up partnerships with the communities affected by TB, for starters. One criticism of the move back to basic biology is that the big funders, vaccine developers and research bodies represent predominantly high-income countries, potentially leaving TB-affected communities without a voice.

But enough of the politics (for now, at least). The fighting that I'm more interested in is taking place on a much smaller scale. I'm talking about what happens when bacteria encounter the cells of the human immune system. One-to-one, *mano a mano*. Because it's here where a vaccine has to have its effect, priming the immune system to respond faster, stronger and better when it encounters an infection. These first interactions decide the outcome of exposure to the pathogen: whether a person will remain healthy, develop active TB or find themselves with a latent infection. Everything starts with the inhalation of bacteria produced by an infected person's coughing or sneezing. In the lungs, these bacteria are engulfed by cells of the innate immune system called macrophages. The innate immune system is the body's first line of defence against an invading pathogen and comprises a number of types of cell, including those that will ingest and digest anything they don't like the look of, Pac-Man-style, then alert the rest of the immune system. If you look on YouTube, there are some highly amusing videos of macrophages chasing and 'eating' bacteria, set to the *Benny Hill* theme tune. I strongly recommend them to anyone who is struggling to visualise what happens to pathogens when they're detected by the innate immune system. To this day, I still picture phagocytic cells behaving like this, complete with the music. Is it scientifically accurate? Who cares?

So, the innate immune system is going to do its best to neutralise the threat. In some people, it will be successful. In 1968 there was an outbreak of TB on the USS *Richard E. Byrd*, in which 66 sailors shared a cabin with seven others who had pulmonary TB. From what I gather, ship cabins are a little like office storage rooms, only with people on the shelves and all their belongings crammed into the filing cabinets. Triple-decker bunkbeds, lots of strapping young men, communal bathrooms. Now multiply that by six

months, and it seems highly unlikely that any of those 66
sailors failed to come into contact with *M. tuberculosis*. Yet in
13 of them, their immune systems had no memory of being
exposed. Their adaptive immune systems – the part that can
remember previous infections – hadn't even come into play.
Exact numbers are hard to estimate, but it's likely that less
than 1 in 10 people have such efficient innate immune
systems. In most people, the adaptive immune system has to
join the melee. This line of defence is slower but cleverer than
the innate immune system. It's also the point at which a
vaccine traditionally gets involved, helping to create those
memories that ensure a quick response should the innate
immune system be breached. We know that the cell-mediated
branch of the adaptive immune system is of vital importance
when it comes to fighting off TB. The issue here is that much
of our knowledge about what protects a person from TB
comes from outliers – people or laboratory animals with a
known defect in a specific part of their immune system. HIV
patients are one such example. Their depleted CD4+ T-cells
(among other things) and the loss of the associated
inflammatory pathways put them at risk of serious forms of
TB, such as disseminated disease. But this doesn't mean that
stimulating a super-strong inflammatory immune response is
necessarily a good vaccine strategy, as poor Robert Koch
discovered when he attempted to create the first-ever TB
vaccine.

Welcome to 1890. My 14-year-old counterparts are
flouncing around in unflattering bustles and frills, eating
calves' feet and swooning over Gilbert and Sullivan. Hopefully
their attractively pale skin and penchant for fainting aren't
symptoms of TB. But with one in seven Europeans dying
of the disease, the odds aren't brilliant. The continent is
overflowing with consumption, as the disease is known, and
the Grim Reaper is a regular visitor to every household. It's
death on a scale that I struggle to comprehend. During the
first half of 2016, the tally of famous dead included David
Bowie, Prince, Muhammad Ali, Terry Wogan and Alan
Rickman. Each of these deaths was greeted as if it were the

exception rather than the norm. As if death were so far removed from our everyday lives that it comes as a surprise when it rears it skeletal head. But in 1890, death was a constant. If it wasn't TB, there were dozens of other diseases or accidents that could claim a person, young or old. That's not to say that death was something to be accepted during the nineteenth century, even if it was somewhat expected. So when the granddaddy of microbiology and discoverer of *M. tuberculosis*, Robert Koch, hinted that he might just have found a cure for TB, Berlin suddenly became the number one holiday destination for the consumptive hordes. They arrived by the trainload, many of them dying on the way or soon after. All of them desperate for this tiny piece of hope. For anyone who is interested, the story of Koch's cure for TB and the overlapping tale of Arthur Conan Doyle's creation of Sherlock Holmes is told beautifully in Thomas Goetz's book, *The Remedy*.

I'll tell just the short version of the story here. Despite the excitement surrounding his miraculous remedy, Koch was particularly secretive about its composition and how it supposedly worked. Quite possibly because he didn't know. The preparation was an extract of dead *M. tuberculosis,* which isn't too strange a concept. Polio, influenza, typhoid, cholera and plague can all be prevented using an inactivated vaccine – dead versions of the disease-causing pathogen. Koch, though, wasn't just hoping to vaccinate against TB, but to treat the disease with his 'vaccine therapy'. He believed that tuberculin, as his remedy was known, initiated the breakdown of infected tissue, effectively starving the bacteria. This is where things get a bit scientifically iffy. While Koch was a trailblazer when it came to much of his scientific work, this idea was a little … wrong. Despite knowing that he hadn't fully researched his cure, nor tested it thoroughly, Koch went ahead with announcing his remedy to much fanfare in an auditorium made up to look like the Temple of Zeus. The lay press were overjoyed, quickly inciting the invasion of Berlin with consumptives. Scientists were not so sure. But this was Robert Koch! Surely he knew what he was doing? (No, he didn't.) I can't help but feel that, after a few decades of success, Koch had perhaps grown a little too complacent when it came to his own indisputable cleverness. A combination of over excitement and misplaced self-belief, perhaps.

The fact that the 47-year-old Koch had used his 16-year-old paramour, Hedwig Freiberg, as one of his test subjects certainly suggests either an over-confidence in his science or under-confidence in his choice of fiancé. In her memoirs, Hedwig says her husband-to-be warned her that she might 'possibly get quite sick' but was 'not likely to die'. Not *likely*. I don't know about Hedwig, but I don't find Koch's phrasing all that reassuring. She went through with it anyway. I related this demonstration of love and trust to my own Hedwig (albeit a male and close-to-my-age version). I was met by a derisive snort of laughter. 'Get lost, you're not experimenting on me,' he said. True love. Anyway, after tuberculin cured – or at least didn't kill – some guinea pigs and Hedwig, it was unleashed upon an excited public. Koch was a star. And then, the first reports of 'problems' began appearing in the medical literature and press. In 1891, a report summarising clinical studies of tuberculin revealed that fewer than 20 per cent of patients treated were substantially improved and more patients died than were cured. The tide turned for tuberculin as the seriousness of its toxicity became apparent. Koch had always been aware that there were side effects – hence the not very reassuring reassurances he made to young Hedwig. But in all the excitement, this message had been lost. Until people started dying, of course.

These side effects, at their best, involved inflammation and pain at the injection site. At worst, tuberculin appeared to make a patient's TB more severe. In fact, it seemed that tuberculin was reactivating latent infections in some patients. What we now know was happening was that tuberculin was over-stimulating the patient's immune system, almost like an allergic reaction, and leading to massive inflammation. The immune response to TB is undoubtedly vital in fighting the disease, but there can be too much of a good thing. Koch's great career was permanently marred by this momentary lapse in scientific judgement, and interest in tuberculin as a therapy waned. Today, Koch's remedy is still used to determine who has been previously exposed to TB, as these people will exhibit a strong skin reaction to tuberculin. When I started working with TB, my occupational health department insisted on firstly checking I'd received the BCG vaccine and then administering me with a Mantoux test. This test involves injecting a small amount of tuberculin beneath the skin and then measuring the size of the swelling. Anything over 15mm ($^{1/2}$in) indicates infection with *M. tuberculosis*. Mine was smaller but very itchy, like a giant bug bite. I'd been previously sensitised by the BCG vaccination, so I expected to experience a small amount of inflammation. If there had been *M. tuberculosis* lurking in my body, my immune system would have been on high alert and would have gone nuts. I've seen pictures of positive Mantouxs and they can look like hideous bubbles of inflammation.

Tuberculin's uneasy journey into the world reminds me of a condition seen in recently diagnosed HIV patients. Immune reconstitution inflammatory syndrome, or IRIS, is the medical equivalent of pretending to lose a fight and then punching your opponent in the back of the head. A patient arrives at the clinic with their immune system in tatters thanks to HIV. So they're prescribed one of the greatest medical advances in modern times – anti-retroviral drugs. And their immune system kicks back into action. Woohoo! Only, all along, a TB infection was bubbling away in the patient's lungs. And paradoxically, when their immune system starts to work again, their TB goes into overdrive and they get really, really sick. Their lungs show signs of massive inflammation. They develop a high fever and start to cough. These traditional symptoms of TB suddenly appear where previously they'd been missing. Inside the body, TB lesions can grow in size and cavitate. IRIS is not often fatal, but can be in more serious cases, particularly those involving TB meningitis. How is it possible that someone with a functioning immune system gets sicker than someone with AIDS?

The answer lies in the observation that HIV-positive TB patients don't transmit their infection as readily as HIV-negative individuals. Transmission is greatly aided by what's called cavitation of the lungs, where infected tissue dies and breaks down to form a semi-liquid material full of bacteria that can be coughed up. But this cavitation doesn't always happen in HIV patients. It turns out that the TB bacillus *needs* a functioning immune system for the infection to progress. This seems counter-intuitive considering how the world is full of pathogens that have spent their entire evolutionary trajectory attempting to find ways to evade being noticed by the human immune system. Take the cold virus, for example. As I write this, I can hear my sleeping toddler over the baby monitor making a noise not unlike a drowning warthog. She is currently suffering from her one-thousand-and-thirty-sixth cold virus of the year (no exaggeration). It's disgusting. Part of the problem is her obsession with licking things. Floors, door handles, other children. But something called antigenic variation is also playing a role. It's the cold virus's version of putting on a silly wig and hoping no one recognises it.

In antigenic variation, the parts of a pathogen that tend to be spotted by the immune system (often the spiky bits on its surface) change over time. Like how famous people seem to change their hairstyle far more frequently than the rest of us. When you're constantly on display, staying the same is a risky choice. So natural selection encourages appearance-changing mutations in the genes of highly recognisable proteins. Over time, a virus tries out lots of different new looks in a process known as diversifying selection while, on the inside, things remain roughly the same. A couple of years ago, Sébastien Gagneux's lab attempted to demonstrate that this process of diversifying selection occurs in *M. tuberculosis*. They took strains from all around the world and compared known antigen genes (Antigen 85A, for example). What they found was that antigens appear to be hyper-conserved. They differ as little as, or perhaps even less than, essential components of the bacterium that absolutely can't change, as then they wouldn't work. That's the opposite of diversifying selection. It's as if natural selection is ensuring that the antigens remain the same to the point that any accidental deviations are lost from the gene pool.

This hints at a worrying conclusion: that *M. tuberculosis* wants to be recognised by the immune system. It's not bothering to change its appearance as it's not trying to hide. It's practising immune subversion not immune evasion, using the host's own defence systems to help it along the way in transmitting to other hosts. Doesn't this raise some tricky questions for the development of effective TB vaccines, I wondered to myself? We all know that the point of a vaccine is to prime the immune system to respond faster in the event that it's exposed to a pathogen. But when it comes to *M. tuberculosis*, it all gets into Goldilocks territory. While an inflammatory response is certainly necessary to prevent disseminated disease, it's not sufficient to prevent pulmonary TB. Then you have the fact that too much inflammation appears to help the transmission of *M. tuberculosis*. Ten per cent of those infected with TB clear the bacteria through the actions of their innate immune system. The adaptive system, though? Based on what I thought I knew about the adaptive immune response to TB, at best, it succeeds in keeping the infection confined to the lungs, in a latent state that can reactivate at any point. I don't know about you, but to me it seems like quite a low bar when it comes to a vaccine strategy. *Hey, let's defeat the biggest infectious disease on Earth by reproducing an immune response that doesn't actually kill the bacteria.* I put this to Harvard TB researcher Sarah Fortune, only less aggressively. 'I've become optimistic about TB vaccine development,' she told me. 'The immune response does a great job at many sites of infection and it's different from the view I had when I came into the field, which was that the immune response does a fine job of controlling but never eliminating the infection. Most sites of infection can be fully sterilised, which says that it's biologically possible. So you just have to figure out how to do it better.' It seems that I have underestimated the natural immune response after all. The issue is in knowing which parts of it are the ones worth replicating.

This idea of picking the 'right' part of the immune response and improving on it is a fascinating one. It's behind an approach I particularly like, which is to attempt to recreate the type of immune response responsible for keeping those strapping young sailors aboard the USS *Richard E. Byrd* free of TB. Traditionally, vaccine strategies have overlooked the role of the innate immune response. The general view was always that the innate system is static and incapable of learning. Then came a Brazilian study of post-neonatal deaths from pneumonia during the years 1986 and 1987. The authors noticed that babies under one year of age who had received the BCG vaccination appeared to have a 50 per cent lower pneumonia death rate than unvaccinated children. It's not just pneumonia. A number of studies have suggested that BCG protects against childhood mortality resulting from a number of infections. Brazilian immunologist Mihai Netea is like Cesar Millan the dog whisperer, only he works with the innate immune system. Where everyone else used to believe the innate immune system to be incapable of learning new tricks, Mihai has demonstrated that it can in fact be trained. A couple of years ago, his laboratory investigated the effects of BCG vaccination on the innate immune system's function. In mice lacking both the cell-mediated and antibody branches of the adaptive immune system, BCG vaccination still protected them against disseminated candidiasis. This effect could only have come from an enhanced function of the innate immune system. Mihai refers to this as 'trained immunity'. Trained immunity isn't quite like the adaptive immune system's memory. It's far less specific, offering cross-protection against a range of pathogens. Mihai believes that BCG vaccination has two effects: one on the cell-mediated branch of the adaptive system, resulting in memory and protection against TB; and one on the innate immune system, inducing cross-protective effects against TB and a range of other infections. The latter effect is faster – low birth-weight babies vaccinated with BCG show signs of protection against infant mortality after just three days.

A number of vaccine researchers are now interested in the idea of targeting innate cells, or cells that straddle the innate and adaptive facets of the immune system. It's in fact becoming increasingly clear that the early interactions of M. tuberculosis and the innate immune system are crucial in determining long-term outcomes of an infection. Modify these early events with a vaccine and perhaps we can modify how the infection progresses as a whole. I'm not going to talk about specific vaccine strategies because many are at an early stage of development and it's impossible to predict what is going to work. What I will mention instead is the idea of delivering a vaccine directly into the lungs. Helen McShane has been using MVA85A to set up new techniques for delivering vaccines via aerosols – like a mechanised cough rather than a needle in the thigh or arm. The hope is that by following the natural route of infection, a vaccine will stimulate the right bits and pieces of the immune system, whatever they are. Immune cells differ depending on where they're located in the body, giving them the ability to shape the immune response in a direction best suited to the local environment. For example, if you look at the types of immune cells present in a healthy lung, you'll find that 90 per cent are macrophages and maybe 5 to 10 per cent are T-cells. In comparison, the immune cells in the blood comprise mainly T-cells and neutrophils. On top of this, the cells of the lung are set up a little differently – while they're primed to switch on signalling pathways that best equip them to fight lung infections, blood cells aren't expecting to bump into a respiratory pathogen. Administering a vaccine to the lungs would hopefully act as a practice run for the immune system. Then, if the lungs ever meet actual M. tuberculosis bacilli, they will be ready. Helen's initial results look promising, in that the aerosol vaccine appears to be safe and induces a stronger lung-based immune response than injection, all without losing the immune response in the blood. Aerosol vaccination is something that's been investigated by the WHO as part of their measles programme. It's likely more suitable than injection as a method for mass immunisations, making it a real possibility for the future.

Overall, there's room for cautious optimism. The vaccine pipeline is beginning to look better than ever and the pendulum swing seems to be returning to a happy medium between basic biology and clinical trials. I'm hopeful that everyone will find a way to integrate all the exciting new ideas to form a cohesive whole. Meeting those ambitious End TB Strategy 2035 targets relies in no small part on finding a new vaccine and finding it quickly. I asked Helen McShane if she was hopeful that a new vaccine would be available by 2025, in time for it to make a difference before 2035 rolls around. '2025 is nine years away,' she said, matter-of-factly. 'I'll be surprised if we have a licensed, deployed vaccine in nine years.' She then went on to tone down her own pessimism. 'We started the [MVA85A] clinical trial programme in 2001, and at the time there were no other vaccines in clinical trials. In the last 15 years I think the field in general has made enormous progress after really what's been decades of neglect, and I think what's important is to continue that momentum.'

Funding needs to increase for this to happen – compare the TB field to HIV if you want an example of what money and determination can achieve. 'We've gone from HIV being a terminal disease with no treatment whatsoever – I used to have a ward full of AIDS patients who were all dying – to it being an outpatient speciality and a chronic disease in which people have a normal life expectancy,' Helen told me. 'If TB had had half the investment that HIV has had, we wouldn't be where we are today.'

CHAPTER SIX

The Human Universe

I'm in Cape Town for my sister's wedding, and I've contracted the Coldzilla of cold viruses. There are more than 200 different cold-causing viruses in the world, but this one is the worst. In my feverous state of misery, I agree to a sightseeing tour. We trundle past the southern slopes of Table Mountain, the Blue Mini Peninsula Tour open-topped bus dragging me through some of the prettiest scenery in the world. Lush greenery, blue skies, vineyards stretching all the way to the jagged horizon. It's idyllic, or so I'm told. Then the bus stops next to a rainbow-hued elephant. The recorded audio tells me that this is the entrance to the Imizamo Yethu township. From the top deck, it looks like a town made of paper. Colourful squares of blue, turquoise, orange and green are, when I look closer, metal containers turned into tiny houses or shacks constructed from a patchwork of wooden planks, corrugated iron and concrete. Lines of drying clothes hang like bunting and water runs down the sides of the streets. Later, I discover that there's no real sewage system in Imizamo Yethu, despite this township being home to around 30,000 people.

Half of the bus hops off to join one of the walking tours promising to give them a taste of the township's culture and vibe, DSLRs already snapping away. We stay put. Partly because I, literally, cannot stand up. Mostly, though, it's because I feel uncomfortable about the idea of gawping at the locals like they're some kind of tourist attraction. The following week, I'm still heavily diseased and suspecting that I have the flu, not just some common cold. I briefly emerge from a Lemsip-fuelled haze to discover that my family has relocated me to the Kruger National Park, where I find myself photographing lions and elephants from our air-conditioned four-by-four hire car. I keep thinking about Imizamo Yethu. I can't decide whether township tours are a bit like a human safari or a profitable way for local communities to get their voices heard. For me, there's a narrow space to navigate between the extremes of mawkish voyeurism and the romanticisation of poverty. The P-word. It was never going to be possible to avoid writing about poverty, as it goes hand in hand with TB. The disease disproportionately affects those who are least equipped to cope with the problem, with poor and marginalised populations bearing the brunt of the epidemic. South African townships are a perfect storm of conditions that provide the ideal environment for TB to flourish. Crowded, unsanitary living conditions, poor diet and restricted access to fresh water, and insufficient health services. It reminds me of Chapter 3 and the TB epidemics that hit Europe and North America. In fact, the levels of TB in Cape Town are similar to those in eighteenth-century London. Only this time round, we know so much more about TB and can treat the disease. So why aren't the people here always getting the help they need?

In particular, stigma surrounding the disease has been a key factor in at-risk patients hesitating to seek a diagnosis. A 2012 study focused on eight townships in and around Cape Town and attempted to look into some of the attitudes towards TB held by those living in these communities. One of their biggest findings was that perceptions of TB tend to be heavily associated with filth and dirt. Many respondents blamed the poorest areas of their communities – *ezimbacwini*, which translates as neglected refugee camps. Flies attracted by stagnating waste were seen as spreading the disease; other people thought that contact with this waste, or simply smelling it, might give them TB. Where racial tensions were present, this could spill over into certain groups being blamed for putting everyone else at risk, sometimes because they were considered to prepare their food unhygienically. The paper also highlighted the stigma associated with HIV as a reason why TB sometimes went undiagnosed. TB patients were often judged by the community as being HIV-infected by the community, especially those who had lost large amounts of weight. It's understandable that individuals might delay accessing healthcare and diagnosis if they believe they might suffer discrimination and prejudice within their community, either as a result of having HIV or of being perceived to have it.

For all South Africa's economic growth to become one of the richest countries in Africa, the divide between rich and poor came as a shock to me (a 10-minute trip down the road from Imizamo Yethu, there's a Michelin-starred restaurant). It seems to be a trend common to many countries in sub-Saharan Africa – the rich get richer, the poor get TB. I'm slightly ashamed to admit it, but this brief South African experience of mine is the closest I've come to real-life TB in a high-burden country. My interest and expertise have always lain with the bacteria and not the people. So it's been an eye-opening experience for me to consider that, in the real world, things are not quite as simple as they are in a controlled experiment. Laboratory models of TB most commonly consider the bacteria in isolation. But no human is a sterile canvas waiting for *M. tuberculosis* to stage its attack. One scientist I talked to about epidemiology told me: 'Putting everything in watertight compartments is a great way to get an NIH [National Institute of Health] grant, but it's not the right way to think about the system of diseases. You need a global outlook.'

This is where that influenza infection of mine changed my entire perspective on TB. You see, by the time I returned home from South Africa, what had started as a brush with the flu virus had progressed into a bacterial throat and chest infection that required a course of antibiotics. Influenza is well known for this. It's as if the virus opens a door to infections that wouldn't be able to get a foot in when it comes to a healthy pair of lungs. I'd not previously considered how one disease might influence the outcome of another, but it's kind of obvious now that I think about it. Our bodies are microbial, viral, fungal and parasitic zoos, full to the brim with a multitude of species. Most are harmless, some are pathogenic, others flit between the two depending on circumstances. Many of these species have been with us for a very long time, nudging at that co-evolutionary relationship between humankind and *M. tuberculosis* over millennia. So in trying to understand what has made TB the disease it is today, we need to also consider all the other species it has encountered over the years and continues to bump into on a daily basis. That's what this chapter is about. 'Real-life' TB with all its microbial, viral and parasitic complications. It seems fitting that I begin with the role influenza has played in the history of TB.

In 1918, a pandemic of the H1N1 influenza virus, or 'Spanish flu', spread around the world, infecting 500 million people and resulting in over 50 million deaths – perhaps as many as 5 per cent of the world's population (many of them quite possibly already weakened by the First World War). To put this in perspective, based on today's world population, this is the equivalent of everyone in the United States dying over the course of a few months. Or nearly all of Western Europe. One of the most unusual aspects of this pandemic is that plotting a graph of age versus mortality rates gives a W-shaped curve. Influenza usually kills only the very young and very old, but in this case there was a large blip in the curve corresponding to high death rates for young adults – supposedly the healthiest chunk of the population. If you superimpose the curve over the age distribution for TB during the early twentieth century, the peak in the middle of the W overlaps for both epidemics. An intriguing possibility is that the influenza pandemic's W was the result of TB patients being pushed over the edge by co-infection. If this was the case, it could have implications for how future acute pandemics progress. Several years ago now, the papers were full of doomsday scenarios about how avian flu was going to kill us all at any moment. Any moment now, any moment. Turns out, scientists were slightly premature in predicting the end of the world as we know it. But should their warnings ever become a reality, perhaps world health policy makers should consider those living in high-TB regions (or perhaps countries blighted by poor health in general) first in line for any flu vaccine.

Support for the idea that at least some of those who died during the flu pandemic were already weakened by TB comes from looking at TB death rates from the time period. A spike in TB mortality suggests that it was the toss of a coin that determined whether it would be the flu or TB that killed a co-infected patient. Those weakened by TB either lacked the strength to fight off flu or flu hastened their death from TB. It's a hypothesis supported by modern data. Today, the global burden of influenza comes in at around 3 to 5 million cases of severe illness a year, and up to 500,000 deaths. In a 2009 flu outbreak, 10 per cent of those who died in South Africa were co-infected with TB. A 2015 study went on to demonstrate that there is an increased risk of death among co-infected patients. Potential explanations for this association include TB lung damage impairing the patient's ability to cope with a subsequent viral infection or increasing the severity of the flu. From the opposite perspective, flu can induce a lung environment that increases a person's susceptibility to bacterial infection – this is likely what happened to me when my own South African adventure turned phlegmy. There's evidence that, on the one hand, flu can increase lung inflammation and damage; on the other, it seems to weaken certain defences required to control infections. When it comes to TB co-infection, this could result in either the aggravation of active TB or even the reactivation of a latent infection.

Based on 1918 records of death, we can only speculate about the role of TB in fuelling deaths from the flu, but it seems likely that it played at least some role in determining who survived. In any case, by summer 1919, everyone had either died or developed immunity, and influenza ran out of the high density of new hosts required to support the pandemic. TB stayed put – a testament to the ongoing success of its ability to form latent, chronic infections. But what was strange was that, after the upward spike in TB deaths during the flu pandemic, the mortality rates reversed and embarked on a rapid decline. Sure, TB deaths had already been following a downward trend in the years preceding 1917–1918, but not to this degree. The slope had been shallow and fairly gentle before flu came along; now it was steep and fast. Fewer and fewer people were dying from TB.

A 1922 letter to *Science* by an A. C. Abbott reads:

*From the standpoint of results, advantageous to the race alone,
and disregarding all humane considerations, this may be viewed
as the beneficent influence of a great plague. The least resistant
of the population succumbed, those more resistant and physically
better fitted to survive, did so. The human material thus left is
probably the most promising that has existed for generations,
in so far as the permanent lessening of tuberculosis among
it is concerned; and we expect that the curve for tuberculosis
death rates in the future will be for a time much more sharply
downward than ever before, [...] For the anti-tuberculosis worker,
the present appears to offer a golden opportunity.*

I can't say I agree with A. C. Abbott's comments in their
entirety. I don't believe it's ever possible to disregard all
humane considerations when it comes to disease and death.
Otherwise we may as well be microbes evolving away in a
culture flask, not caring who lives and dies just so long as the
strongest genes persist in the population. From a scientific
standpoint, however, Abbott raises an interesting theory
about how disease can shape human evolution and impact
upon other diseases. It's an idea that has been taken up
by Andrew Noymer of the University of California, who
believes that selective mortality during the influenza pandemic
hastened the decline of TB in the US. He describes the
pandemic as a pivot point for TB rates, setting them on a
faster downward trajectory towards the present day. 'You can
only die once,' he told me. 'You can think of the tuberculous
population of people in 1918 as a bathtub that's simultaneously
filling up with water from the faucet and draining at the same
time. In most years back then it was draining a bit more than
it was filling. What the flu epidemic did was it emptied the
bathtub with a bucket to reduce the size of the population
with TB.' The 1918 pandemic 'saved' around 500,000 people
from dying of TB – first by killing in the order of hundreds
of thousands of co-infected individuals, and second by having
a knock-on effect in removing a large number of infectious

people from the population as well as their potential victims, therefore slowing transmission. It's an effect that lasted some 10 years.

This idea of an overlap between unrelated diseases is an interesting one. TB and flu are both diseases of the lung, so it's easy to understand why one would worsen the other. But what about other diseases that don't even physically meet in the body? One of TB's rivals for the dubious title of world's oldest and most deadly disease is malaria. A study led by Carsten Pusch of the University of Tübingen attempted to use ancient DNA to look for traces of both malaria and TB in 16 mummified heads (sans bodies, for some reason) from the necropolis at Abusir el-Meleq, dating as far back as 800 BC. They found high frequencies of malaria and TB co-infection (25 per cent) in the pre-adolescent and young adult severed heads. Fast forward to today and malaria-TB co-infections are no longer an issue in Egypt. Cross the Sahara Desert, though, and we find ourselves back in sub-Saharan Africa. It's a big area of the world, encompassing everything from dusty villages to polluted cities. But if we look at a heat map of the worldwide incidence of TB and superimpose it over a map of malaria, there's a big red patch focused on most of sub-Saharan Africa. I'd always known this but, surprisingly, had never stopped to consider what the overlap between these two global killers might mean for TB. And it looks like I'm not the only one.

There are just a handful of papers looking into malaria-TB co-infection. One of these studies was based in Angola, just a couple of countries up from South Africa. The capital, Luanda, is a scarred city marred by sprawling slums and chaotic traffic jams coloured blue and white by Candongueiro taxis. It's also one of the most expensive places in the world. The estimated cost of living is 26 per cent higher than London and 20 per cent higher than New York. Here, the gulf between the oil-fuelled rich and the poorest of the poor is as extreme as it gets. The beach front is a glittering jumble of skyscrapers, white sand, palm trees and dozens of towering cranes. The slums are seas of corrugated metal, concrete and litter; a remnant of the 27-year-long civil war that saw millions of Angolans take refuge in the city. In 2013, a team from the University of Lisbon collaborated with the city's TB hospital – Hospital Sanatório de Luanda – to look at the problem of co-infection in TB patients. They found that more than one-third of the patients were co-infected with malaria and 37 per cent co-infected with HIV. With such high levels of malaria-TB, it's even more surprising that so little is known about how the two diseases interact.

A paper from Bianca Schneider's lab in Germany used a mouse model to look at what happens to the immune response to both diseases upon co-infection. Mice with chronic TB exhibited worsened symptoms when the malaria parasite was added to the mix. *M. tuberculosis* replication ran away with itself and the lungs were infiltrated by inflammatory immune cells, worsening the tissue damage. The problem here is that both pathogens stimulate inflammatory branches of the immune system, potentially causing what's known as a cytokine storm. This is similar to what happens in sepsis when the body basically loses control of its own immune response. In addition, malaria infection suppresses key parts of the immune response important to TB control. So you now have a situation in which the response to *M. tuberculosis* is both disproportionately strong and inflammatory, but also insufficient to control bacterial replication. Conversely, TB infection provided the mice with some protection against malaria. The co-infected mice had reduced parasite levels, less liver damage and lower weight loss. The likely explanation here is that the inflammatory pathways already switched on by an existing TB infection mean that the immune system is ready to deal with a new malaria infection straight away. But what does this mean for the real-life management of this co-infection in humans?

If you drive about 7,000km (4,500 miles) out of Luanda along the west coast of Africa, you reach the country of Guinea-Bissau. It's a beautiful place, with lush green forests, white sandy beaches and blue waters. It's also right up there among the world's poorest countries, with the capital city of

Bissau still bearing the marks of civil war in the form of cracked roads and poor living conditions for the impoverished majority. The Hospital Raoul Follereau in Bissau is the go-to place for those with severe TB, referred there from hospitals all around the country. It's not a good sign to be sent to the Hospital Raoul Follereau. In the 2004 rainy season, the mortality rate at the hospital peaked at 60 per cent of patients. Even with its specialised focus on TB, the patients sent there were still dying at an alarming rate, and according to the hospital's physicians, most of the deaths were from malaria, not TB. It seemed that patients already severely ill with TB and possibly also suffering from the effects of malnutrition or HIV infection were particularly susceptible to contracting the most serious forms of malaria. A team led by Italian Fabio Riccardi wanted to see if preventing malaria could improve mortality rates among TB patients. Each patient was provided with an insecticide-treated bed net, the hospital was disinfected regularly, health education was provided to teach everyone how malaria can be prevented and patients were all given preventative antimalarial drugs. The mortality rates dropped by 7.5 per cent, and all for less than €2 per person. Such simple, cheap intervention methods aren't going to solve the problem, but they're a step in the right direction.

What all of this does highlight is that attempting to combat one infection without considering coincident pathogens would be very foolish. Both the examples above are acute diseases with relatively short-lived effects. The overlap between TB and malaria, or TB and influenza, is a brief one. Chronic diseases, on the other hand, have the potential to work together at a deeper level. The next example I want to mention isn't as deadly as malaria but is far more widespread. Up to 2 billion people worldwide are infected with helminths – or worms, to use their wrigglier name. Part of the helminthic survival strategy relies on immunomodulation – shoving the immune response in a direction that makes the host a more worm-friendly environment. It's a strategy honed over hundreds of millions of years of co-evolution with the worms' vertebrate hosts, the end product of which is the ability to form a long-term, chronic infection that rarely kills on its own. Despite this, however, helminth infections do cause a huge amount of disability and suffering across the world, and contribute to more than 100,000 deaths (the total is likely much higher due to the link between helminth infection and malnutrition). The most highly affected areas of the world yet again overlap with those bearing the greatest TB burden, meaning helminth-mediated immune subversion may be impacting upon the response to *M. tuberculosis*. Long-term worm infection, you see, inhibits the inflammatory branches of the immune system required for the initial response to a TB infection.

This dampening down of the immune response to TB has implications for the vaccination and diagnosis of the disease. The areas of the world where vaccination against TB (the BCG vaccine) works the most poorly also happen to be the regions where helminth infection is endemic. There's some evidence that unborn babies can become sensitised to helminth antigens passed to them by their mother and that this exposure can have long-lasting consequences for immunological memory, effectively setting up their immune system with a skew towards anti-inflammatory responses. Even if a baby is given BCG within the first hours of life, it's possible that its immune system won't react to the vaccine with the type of response required for immunity. That said, a large Ugandan study treating mothers with anthelmintic drugs led to no improvement in infant BCG vaccination responses. So the question of whether helminths modulate the response to BCG vaccination remains up in the air, but is something that should be considered for future vaccination strategies.

The research into whether helminth infection can worsen active TB is even less clear-cut. A study based in north-west Ethiopia looked at the incidence of worm infections in TB patients compared to people who'd been exposed to TB but had remained disease free. They discovered intestinal helminths in 71 per cent of the TB patients but only 36 per cent of the controls. But while one study suggested that co-infection resulted in more advanced TB, another concluded that helminths may actually be beneficial when it comes to controlling the overall burden of *M. tuberculosis* in the human lung. A recent Ethiopian trial also failed to improve TB outcomes using anti-helmintic drugs to remove the worms from the co-infection equation.

What interests me most about helminth-TB co-infection is the idea that TB evolved in humans who were likely inhabited by a number of other pathogenic species. There's an old episode of *The Simpsons* ('The Mansion Family', episode 12, season 11) in which Mr Burns visits the Mayo Clinic and discovers that he has all known diseases as well as a few undiscovered ones. The balance of diseases in his body keeps him from dying, presumably because they all interact in ways that prevent one from taking him out. Despite the doctors warning him that even a slight breeze could disrupt the balance, Mr Burns decides that he is indestructible. What I love about this is that reality does, in some cases, mirror *The Simpsons*, believe it or not. People infected with the stomach ulcer bacterium *Helicobacter pylori* appear to be less likely to develop TB. *H. pylori* has been colonising people for at least 50,000 years and, with *M. tuberculosis*, is among the most common pathogens in the world. But, like helminth infection, its prevalence in the Western world has decreased over time. Is this due to increased antimicrobial use, better sanitation and less crowding? Or is it that *H. pylori* provides a survival benefit against diseases that disproportionately affect those living in low- and middle-income countries? You've got to wonder about two infectious agents that can survive together in the human host for an entire lifetime, both chatting away to the immune system to ensure their own survival. Could the memory of each one be etched in the other's genes, just as *M. tuberculosis* has left a lasting impression on the human genome?

A study led by Julie Parsonnet looked at people who had been in contact with TB patients from Karachi in Pakistan and from The Gambia. More than 60 per cent were positive for *H. pylori* infection, and these individuals were around half as likely to develop TB after close contact with an infected person. The authors of the study propose two ways in which

H. pylori might protect against TB infection. The first is the hygiene hypothesis that you may have heard of in conjunction with the development of allergy. You know – the excuse that some of us (ahem) use to console ourselves when our child eats food off the pavement or we don't bother cleaning the house for two weeks. When it comes to TB-*H. pylori*, the theory is that early infection with *H. pylori* trains the immune system to respond in a specific way to future challenges. It could be that it happens to push the immune system in a direction that makes it better at dealing with *M. tuberculosis*. This would be similar to the proposed link between decreasing helminth infections in high-income countries and the rise of autoimmune disease. Or *H. pylori* could induce what's known as bystander effects, in which it keeps the immune system on high alert. This could theoretically prevent *M. tuberculosis* from slipping under the radar when it arrives on the scene.

There are some questions surrounding whether *H. pylori* is friend or foe. On one side, you have the fact that it potentially provides some protection against a number of diseases, including oesophageal cancer, severe gastroesophageal reflux disease and maybe even asthma. On the other, there's a raised risk of peptic ulcers and gastric cancer. I find it hard to think of *H. pylori* as a true commensal species, even though it infects something like 50 per cent of the world's population. True commensals, in my book, don't routinely make people ill, so perhaps a better way of viewing it is as a highly successful pathogen. Either way, it does bring me nicely round to all the non-pathogenic species living in and on us. Based on a recent count, our bodies are only half-human, with something like 30 trillion human cells plus 39 trillion bacteria, most of which live in our guts. The gut microbiome is, in fact, one of the densest ecosystems on the planet, with something like 10^{11} microbes per millilitre of, um, large intestine colonic content. As my human 'scaffold' types away at this chapter, all its microbial passengers are getting on with the tireless task of looking after my insides. Among its many functions, the gut microbiome plays a critical role in training our immune system. *Yes, attack that pathogen. Kill, kill, kill. No, that's not your enemy, leave it alone!* It's *Rocky III* on a microscopic scale (he was nothing without his trainer Mickey).

The gut microbiome has long arms (pili) and its effects reach all the way to the lungs, among other places. For example, there's an association between the types of bacteria found in the gut and the development of asthma. Basically, it's possible to read the contents of a baby's nappy as a kind of poo-runes scenario, and use this to predict whether that baby will go on to develop asthma at a later date. This got me thinking about whether there's a link between TB and the gut microbiome. Surprisingly, though, there's not much research out there, despite the importance of the gut for a person's immune system, and the importance of the immune system for fighting TB (among other things). But I did find one 2014 paper from William Bishai's group in the US looking at what pulmonary TB does to the mouse gut microbiome. Upon infection, the mice exhibited a rapid decrease in gut bacteria diversity and big shifts in community structure. The presence of a TB infection in their lungs was able to upend the gut microbiome despite the two populations of bacteria – lung and gut – never physically meeting. Bishai's team are no longer working on this project, so I don't want to over-emphasise their conclusions when no one else has published anything along the same lines. I do, however, like the idea that not only does the immune system take its cues from the gut microbiome, but the immune system also talks back and influences what species can call our intestines home. The way I think about it is that a TB infection is activating the immune system in such a way that the environment in the gut is altered and the composition of species changes as a kind of collateral damage.

I was ready to leave this section there until I came across the concept of the lung microbiome. This confused me somewhat. You see, I'd read the textbooks stating that the lungs are sterile. Sterile, microbe-free, clean as a freshly bleached toilet seat. No microbes here. Unless, of course, something has gone horribly wrong and you've contracted an infection such as TB. That's right, isn't it? Actually, no. This has to be an example of lots of people not stopping to question what is actually quite a silly belief when you think about it. Because, come on, we've found microbes buried beneath sheets of ice in Antarctica, yet we're saying that the lungs – a nice, warm, moist environment just a few inches down from the germ-ridden mouth – are incapable of supporting life? Thankfully the scientific community has coughed up this rather silly presumption and it's now accepted that, like nearly every other environment on earth, the lungs have their own microbial ecosystem. Sure, healthy lungs contain a very low level of bacteria (and I'm only going to talk about the bacteria, as so little is known about viruses and fungi). It's not on the level of the gut microbiome, mainly because the lungs aren't full of a nice constant supply of food. But bacteria are definitely there, even if they're really difficult to study.

As the field of lung microbiome research grows, it's becoming increasingly obvious that disease alters the complement of bacteria living in the airways. I can count the number of TB-lung microbiome papers on my fingers, but the majority of them come to the same conclusions. That TB causes a shift in the lung population structure and increases species diversity. *M. tuberculosis* not only interferes with the host immune response but also causes physical damage to the lung tissue. This could make it easier for opportunistic bacteria to get a pili-hold. These species may in turn help *M. tuberculosis* cause further lung damage, playing their own part in the progression of the disease. You can think of the TB-infected lung as a unique ecological niche that can play home to a range of bacteria that would normally not be found in the airways. It reminds me of the London riots of 2011. It was as if it suddenly became acceptable to loot and smash things up, and the violence spread from borough to borough, city to city. Some of those involved were already angry with the system at large and took advantage of the situation to unleash their frustrations; others got caught up in the 'well, everyone else is doing it' mentality and momentarily forgot that acting like an arsehole isn't a good life choice. For a while, parts of London were overrun with idiots, until the police brought everything back under control and I could once again pretend that my fellow humans aren't just one smashed window away from a complete breakdown of social morality. In the lung, TB is the one inciting riots then sitting back to take advantage of the chaos.

What interests me about the TB lung microbiome work isn't just its potential to tell us more about how *M. tuberculosis* interacts with the human host and its complement of microbes, but the possibility that it might answer some of the big questions in TB research. Such as why do some people harbour a latent infection all their life while others progress to active TB, and why does drug treatment work in some people and not others? So far, there's no evidence that the complement of lung (or gut) species influences disease outcome. But even if it is found that the microbiota is merely a fingerprint of a patient's disease status rather than a contributing factor, surely it could be used as a marker of those who are in need of medical intervention? A couple of years ago, a Chinese group based at Fudan University looked at the microbiota found in coughed-up TB sputum. Like previous studies, there were big differences between the TB patient microbiomes and those of healthy controls. There were also differences between the recurrent and new TB patients, and between those whose treatment had failed and those who were cured. The authors suggested that the presence of certain bacteria may be associated not only with the onset of active TB but also with the recurrence of an old infection and treatment failure. Diagnosing someone based on their microbiome is not a new idea. Variations in gut microbe ratios can predict hospitalisations in diabetes patients, complications after colon surgery, the success of stem cell transplantation, the likelihood of transplant rejection and the future development of obesity in children, to name just a few. Perhaps, one day, we'll be able to put our microbiome to use in predicting our likelihood of developing a host of diseases. At the very least, I would like to see medicine move towards a whole-organism approach in which treatment considers not just one species but the whole human universe.

Many of the microbiome species have been with TB since the beginning, likely colonising our distant ancestors back in the Cradle of Life. This wouldn't be a very complete chapter on co-infections, though, if I didn't mention a far newer player in the evolution of TB. The African continent doesn't have the highest incidence of TB in the world. That dubious title goes to South East Asia. What sets sub-Saharan Africa, in particular South Africa, apart from countries such as India and China is that the TB epidemic here is being fuelled by a second infectious agent – HIV. Back in 1991, *The Lancet* published an article titled 'Is Africa Lost?', focusing on the deadly synergy between HIV and TB. It's 2016 as I write this and, as far as I'm aware, Africa is still there, although so are TB and HIV. Let's look at some stats. In 2014, there were 36.9 million people living with HIV/AIDS worldwide, 25.8 million of them in sub-Saharan Africa. Of the 9.6 million new TB cases in 2014, 1.2 million were among HIV-positive individuals, 74 per cent of whom were people living in Africa. In the same year, one in three HIV deaths was due to TB. But it's not all doom and gloom, as these shocking statistics actually represent an improvement on previous years. I'll come back to intervention methods in a later chapter, but I am including these stats here to highlight the sheer scale of the problem and how it's impossible to talk about TB without also mentioning HIV.

The whole is greater than the sum of its parts, with both infections egging the other on and making everything more difficult. This is one of the biggest problems in the South African townships such as Imizamo Yethu. Close to 90 per cent of 30- to 39-year-olds living in these informal settlements have a latent TB infection. With between a quarter and a third of the community carrying the HIV virus, and HIV pushing latent TB towards reactivation, rates of active TB are extremely high. I read that, in one township, there are around 2,000 cases per 100,000 people – that's 2 per cent of the population developing active TB every year. I talked to Alex Pym at the KwaZulu-Natal Research Institute for TB-HIV (K-RITH) about TB-HIV in South Africa. 'What's interesting about South Africa is that in African terms, HIV came late to South Africa relative to Uganda [in] Central Africa,' he said. 'The seroprevalence in the early 1990s in antenatal women was around 2 per cent. At that time, TB incidence was still high but manageable. Then KwaZulu-Natal had this explosive epidemic, and it's now got the highest rates of HIV in southern Africa, if not the rest of the world. The HIV epidemic literally doubled – 2, 4, 8, 16, 32 over the period of a decade.' Over the same period, there was a 10-fold increase in TB rates.

Some of the South African problem stemmed from the government's hesitation to adopt antiretroviral therapy (ART). Back in 2000, when the growing consensus was that ART needed to be made available outside of Europe and America, policy makers instead chose to embrace AIDS denialism. In all, more than 300,000 people died as a result, and the uncontrolled HIV situation contributed to South Africa's current TB burden. In the end, some US-funded ART programmes started to make their way through, but they set the tone for ART clinics to be separate from TB healthcare despite the two infections being so intrinsically linked. Because of the unholy partnership between TB and HIV, it's vital that both infections are diagnosed quickly before a patient gets seriously ill. Diagnosis, unfortunately, is hampered by a number of factors including stigmatisation and, in the case of TB, a lack of reliable and quick diagnostic tests. Both diseases require extensive courses of drugs with sometimes nasty side effects, leading to non-compliance in many patients. And of course, the drugs are expensive and, even when available, place a huge strain on countries affected by the dual pandemic.

There are also issues with drug interactions between ART and TB therapy (the new TB drug bedaquiline, for example, does not mix well with ART), and the lack of clinical trials of new TB regimens in HIV-positive people, despite 70 per cent of KwaZulu-Natal drug-resistant cases occurring in combination with HIV. 'Traditionally, pharmaceutical companies who are designing clinical trials for TB drugs are very risk-averse,' Alex Pym said. 'If you have a study with a very high mortality or a lot of side effects, they may wrongly be attributed to the drug and not the HIV. We need those clinical trials done with HIV-positive patients. The sooner we include HIV-positive patients, the sooner we'll get the data to use those drugs safely.'

Anyway, that's the problem (briefly) from the TB control side of things. What I'm more interested in here is how the two pathogens interact on a biological level. I already talked about how HIV can toss aside the association between certain TB lineages and a patient's racial ancestry, indicating that it's stamping all over millennia of co-evolution. And I've talked about how HIV-mediated immune suppression leads to scarily high latent TB reactivation rates in this population.

These are just two of the ways that HIV has changed the face of TB. Let's continue our trip around the vast African continent and head over to the growing cities of Botswana to look at another effect. The land of the big-game safari, Botswana has rapidly transformed itself from one of the poorest countries in the world to one of the most stable economies in Africa, thanks in the most part to its diamond trade. Once, the country had the world's highest rates of HIV infection, but in recent years has begun to bring the epidemic under control as a result of an advanced treatment programme. Botswana was the base for a 2015 study looking at TB inpatients with increasing levels of immunosuppression due to HIV infection. As HIV progresses, a patient's CD4+ T-cells are diminished (there are other effects that I won't go into). Below 200 cells per millilitre of blood, the immune system is compromised to the point that a patient is diagnosed with AIDS. The researchers found that in those with normal levels of CD4+ T-cells, TB infections tended to be a result of a single strain of the bacterium. However, as the CD4+ T-cells declined to fewer than 100 cells per millilitre, around 20 per cent of patients were infected with more than one strain of *M. tuberculosis*. In patients who had previously been treated for TB, this went up to 41 per cent. This goes against the traditional view of TB being caused by a single strain. But why does that matter, I hear you ask? Because it has implications for how we treat the disease and view the development of drug resistance.

A friend of mine recently worked on the strange case of a London TB patient who, because of his excessive alcohol use, had to be hospitalised in isolation for the duration of his treatment. That way, the medical staff could ensure he took all the medication as directed and take care of him. The strain of *M. tuberculosis* isolated from his lungs was sensitive to first-line drugs and it should have been a simple case to treat. After four months, though, he still had bacteria in his sputum. And when the doctors grew up a new culture, it was resistant to two of the drugs in his four-drug regimen. He'd definitely been taking the right medications. A whole team of doctors had been involved in his care – they'd followed the guidelines to the letter and someone had watched him take every pill. And he couldn't have picked up a new infection, as he'd been isolated in a negative-pressure room for the entire time. The only possibility was that he had been harbouring a mixed infection right from the beginning. It's likely that the drug-resistant strain was there in low numbers, held at bay by the quicker-growing drug-sensitive strain. But once the drugs started to kill the sensitive strain, the resistant one was unmasked and able to take hold. The significance of mixed infections applies on a number of levels. Importantly there are implications for understanding the role of reinfection or relapse in recurrent TB – is it a patient's old infection popping back up, which means it wasn't treated properly in the first place, or is it a new infection, in which case transmission is more of a problem than anticipated?

What this sort of work indicates is that a TB infection is not a simple case of one bacterial strain infecting one human within a void. The relationship between *M. tuberculosis* and humankind is entangled in a web of other interactions. At the same time as finding ways to survive alongside the human immune system, *M. tuberculosis* has learned to coexist with the

other species it encounters during its lifecycle, be they other
pathogens such as helminths, commensal species making up
the gut and lung microbiomes, or something in between like
H. pylori. The real world outside the lab, as it turns out, is a
complicated place, and every new TB infection is a whole
new ecosystem. Piled on top of a person's unique DNA,
there's their microbiome (and the poorly understood virome),
and on top of that other chronic infections that can alter how
the immune system behaves, and on top of that acute
infections such as malaria and influenza that can alter the
course of the disease. There's a whole universe inside us that
potentially influences the outcome of an infection.

Today, TB's biggest partner in crime is HIV, and the virus
is already not only changing the face of TB but also reinventing
many other co-infections, co-morbidities and even the
human microbiome. But what has surprised me the most
about writing this chapter hasn't been all the evidence that
M. tuberculosis is influenced by other species. It's been the
realisation that none of the science really matters as long as
those who are getting sick don't have access to diagnosis and
treatment. For most people, it's not their microbiome or any
co-infections that are really going to determine whether they
are at risk of TB. It's where they live in the world and, even
more importantly, how they live. All those social factors
remain the rainbow-hued elephant in the room that everyone
can see but no one seems to know how to fix.

CHAPTER SEVEN

Huber the Tuber's
20-Tuberculear Sleep

Keep calm and carry on. Nasty von Sputum is waging war on Lungland, slashing his sword around into the delicate air sacs of the Promised Land O'Lung. He's a fiend, all right. A little bacterial bastard. I mean, look at that tiny little moustache perched on his top lip. And his evil deputies, Gorring and Gobbles. And all that Seig heil-ing. I don't know about you, but I'm getting some pretty strong Hitler vibes here. Surely not, though. I mean, no one would really write a book using an anthropomorphic Hitler-microbe to tell the story of how *M. tuberculosis* causes disease. Would they? Harry A. Wilmer *would* – or did, to be more accurate. In 1942, *Huber the Tuber* was published by the National Tuberculosis Association as a kind of public outreach attempt. Finding this book was a funny coincidence for me, actually. Just a day earlier, I was waiting at the doctor's surgery and was lucky enough to have the time to read every single one of the posters trying to convince me I have prostate cancer and childhood meningitis and am currently mid-stroke. There was something missing from them, though. 'You know what these posters need?' I said to my toddler. 'More Hitler ...' (only some of this story is actually true).

The book's real, though, and chronicles the adventures of Huber the Tuber as he and his comrades ride into their new home on some coughed-up droplets of spittle. Once inside the lung, the naughty bacteria set to work carving out a niche for themselves at the expense of the poor host. After a brief interlude for some timeless misogyny that sees Huber's love interest washed away by the bloodstream as 'a horrible example of what happens to adventurous wives, even in Lungland', the tubers engage the host's immune system and 1940s medicine. Battle, in all its phlegmy glory, ensues. Huber, though, is a voice of caution among the bloodthirsty tubers. I don't believe that he's the good guy the book paints him to be (he doesn't appear to mourn his poor dead wife, for starters). No, my theory is that Huber is a coward. Sure, he gets to be the hero of the story, but everyone's a hero in comparison with Hitler. While Hitler-microbe wants to invade, invade, invade, Huber has less ambitious lifegoals that, through pure coincidence, happen to mesh with the owner of the lungs' wish to not die. Huber, you see, believes in playing the long game and using his patience to his advantage. He knows that his best chance of survival is to avoid taking on a healthy immune system aka the Home Guard Army. Huber is a proponent of the latent infection.

'Don't blow a cavity; don't cut the lung up. Everyone can have an airsac to himself,' Huber tells the other Tubers. 'We do not want a big spread; we do not want miliary-military aggression – we want peace and quiescence.' The conservative Tubers build homes from calcium stones and make themselves comfortable, while Huber settles down with a book - The Magic Mountain by Thomas Mann. 'Soon he fell asleep and slept for twenty tuberculears!' the story tells us. Incidentally, *The Magic Mountain* is a looong German book set in a TB sanatorium, starring a rather shallow young man who arrives for a visit and somehow ends up staying for seven years. There's a lot of snow and a lot more philosophising on the nature of the spirit and time, among other things. If I ever find myself hospitalised for TB treatment, I shall take this book with me. If the disease truly is capable of opening up a person's soul, as nineteenth-century attitudes held, then perhaps I will find myself able to finish *The Magic Mountain*. I keep trying but all I achieve is the smallest inkling of what it must be like to lose a big chunk of your life to something like TB. Time that is gone forever and for which you have nothing to show. I can recommend *Huber the Tuber*, though. Because, beneath all the weirdness, it's actually full of fairly accurate science and a real message. The preface starts with the words 'Wars breed tuberculosis', and suddenly all the 1940s joviality reveals something far more serious.

Publications like the whimsical *Huber the Tuber* manage to shine a light on the human toll of what was the leading disease of the day. At the time of its publication, around 50,000 people were dying annually in the US as a result of TB. Of these, 30,000 were young people (15–45 years of age). In the same year, over 20,000 UK deaths were attributed to TB. Poor nutrition, general ill health, cramped living conditions. Thanks, War! I've often heard it said that disease is the only victor of war, and TB certainly takes more than its fair share of the spoils. The end of the Second World War, for example, brought an onslaught of veterans returning from the front line with TB. The civilian toll was equally worrying. In Germany, death rates rose from 60 per 100,000 people before the invasion of Poland to 260 per 100,000 from 1945 to 1946. While London kept things under control and saw a relatively small rise from 60 to 100 per 100,000, Warsaw was at the opposite end of the spectrum, going from 155 to 500 per 100,000. For many of those returning from concentration camps, where as many as 1 in 10 inmates had active TB, the battle was far from over and had the doubly awful result of spreading disease to those back home.

Today, persons affected or displaced by war, political instability and natural disasters remain at high risk of TB thanks to poor living con.ditions and a lack of access to consistent healthcare. Tibetan refugees in India, for example, have rates of TB almost double those of India overall. And Lebanon (among other countries) has reported a 27 per cent increase in TB rates, mostly due to the influx of Syrian refugees. Large numbers of Somalian refugees have crossed into Kenya seeking treatment for TB. Afghanistan has seen an increase in TB rates over the last 10 years. The Democratic Republic of the Congo continues to struggle with TB after years of war. I could probably write an entire book on the problem of TB in vulnerable populations but I suspect you get the idea. As if all that wasn't enough, interrupted treatment in refugee populations means that drug resistance can take hold, as is the case in Ukraine. The problem here can be traced back to the collapse of the Soviet Union. In Thierry Wirth's seminal sequencing paper, he followed two particularly stubborn members of the Beijing Lineage back to the dissolution of the communist state. What this means is that today's multidrug-resistant (MDR) TB situation in parts of Eastern Europe was set in motion more than 20 years ago when the Soviet Union's previously free health system decayed into ruin. On top of that, HIV infection rates rose and inequality between the rich and the poor flourished. Of the 27 high MDR-TB countries as listed by the WHO, 14 are former Soviet states.

Drug resistance in Ukraine had been smouldering away since the Soviet collapse, but the recent military conflict between Russia and eastern Ukraine has fanned the flames. Around 1 million people have been displaced by the fighting, leaving them at high risk of communicable disease. Hospitals have closed or been destroyed and few medical staff remain in the conflict zones. A 2014 correspondence in the *International Journal of Tuberculosis and Lung Disease* suggested that, at the time of writing, there were 5 million people (3,000 with MDR- or XDR-TB*) living in conflict-affected areas without access to medications. From 2009 to 2013, rates of drug-resistant TB tripled, leaving Ukraine with a higher burden of MDR- and XDR-TB than all the European Union countries together. The letter authored by a number of Ukrainian doctors and scientists calls for support from neighbouring partners and international donors to 'prevent a medical catastrophe in the

---

* Extremely drug-resistant (XDR) TB is MDR-TB with additional drug-resistance mutations, making it even harder to treat. Chapter 10 goes into more detail about all the various levels of drug resistance at play.

heart of Europe.' The authors state that Ukraine will not be able to control the problem alone and that drug resistant TB is just one of the medical issues afflicting those caught up in the conflict.

A 2010 Images to Stop Tuberculosis Award was granted to Moldovan-born Misha Friedman for a series of photographs showing the human face of TB in Ukraine. The black-and-white images are hauntingly beautiful. In one, two drawn patients help an elderly woman to her hospital ward, performing the job of healthcare workers. In another, a seriously ill man sits in a cramped TB ward with cracked walls and narrow beds. He has HIV, hepatitis and drug-resistant TB, and is a drug user. War, displacement, homelessness, substance abuse, imprisonment, poverty. It feels like the TB situation in many countries teeters on a precarious edge. Under control but only until something shoves the balance in the bacterium's favour, be it war or the more insidious creep of HIV infection. TB is always there, waiting in the shadows, ready to take advantage of a host's poor health to reactivate a latent infection, or ready to transmit freely thanks to the overcrowding and lack of medical care that walk hand in hand with displaced populations or those affected by war.

I'll come back to the problem of transmission in a later chapter, but for now I want to take a closer look at what *M. tuberculosis* gets up to during a latent infection. It's been estimated that around 2 billion people are asymptomatically infected with TB. Eradicating these bacteria before they can reactivate or predicting those people most at risk of developing disease would defuse the ticking time bomb of latent infection before a spark sets it off. But first we need to understand how the bacterium survives for such long periods of time despite the best efforts of the human immune system.

So, let's take a closer look inside the calcified home where Huber spent his 20 long tuberculears. Following inhalation of infectious droplets, TB bacteria are taken up by macrophages. In most patients, instead of being killed by these immune cells, *M. tuberculosis* starts to replicate. As the infection progresses, macrophages and other immune cells are recruited into an organised structure called the granuloma. In the classic form of the granuloma, the centre breaks down to form a cheese-like substance and the outside of the structure becomes enclosed by a thick, multicellular wall. On a chest X-ray, this type of lesion shows up as a white ring surrounding a black bubble. Other granulomas can look like small white clouds against the black background of healthy lung. In a dissected lung, you might see different types of granuloma as yellowy lumps a bit like embedded cannellini beans or maybe something that looks like the empty husk of a baked bean surrounding a hole in the lung tissue. Active disease can involve multiple granulomas, some of them spilling out bacteria that go on to seed more foci of infection – a chest X-ray can look as if those white clouds are filling an entire lung. In asymptomatic or latent TB, the granulomas are, on average, better controlled. Sometimes, a routine chest X-ray prior to hip-replacement surgery, for example, will pick up a latent TB infection. A person could have been infected for most of their life with no symptoms and, in most cases, will never develop active disease.

The traditional view was that the TB granuloma is a good thing for the host as it contains the infection, effectively walling it off and controlling it. In many cases, the actions of the immune system might sterilise the granuloma, leaving behind no more than a little scar. But in others, *M. tuberculosis* – the Bear Grylls of the microbial world – finds a way to survive. There's an illustration in *Huber the Tuber* showing the hero napping inside his TB lesion with the explanation: 'Once the tubercle bacilli are walled in by calcium they usually remain inactive and cause no trouble. However, we can think of them as sleeping or "playing possum" because under circumstances favourable to the bacilli they may escape and cause the greatest harm!'

For the sake of scientific accuracy, I'll mention that all Harry A. Wilmer's calcification descriptions are possibly not strictly accurate. Calcified granulomas could well be ones that have healed rather than the home to bacteria. There's also been some debate over whether *M. tuberculosis* really stops growing and 'goes to sleep' during a latent infection. An alternative hypothesis is that the low numbers of bacteria present during latent TB are the result of a dynamic equilibrium between bacterial cell division and killing by the host. I personally think it's probably a bit of both – sleep and killing – but I'll start off by focusing on the hibernation side of things, as it's what I used to work on.

So, what is it that sends Huber/*M. tuberculosis* to sleep within the granuloma? The main theory is that granulomas limit the growth of *M. tuberculosis* by cutting off the bacteria's supply of oxygen and nutrients as well as exposing them to various toxic agents. My own work, back when I was still in the lab, focused around the idea that oxygen starvation – known as hypoxia – is intimately linked with latent disease. I know, I know – you wouldn't automatically associate the lungs with a *lack* of oxygen, but bacteria hiding inside a macrophage hidden away in a granuloma are actually pretty much cut off from everything. The hypothesis goes that *M. tuberculosis* can't grow without oxygen, so it enters into a non-growing state to wait for better times to arrive. Some of the evidence to support this idea comes from the observation that a TB infection best takes hold in areas of the lung with the highest oxygen availability. In humans, these are the upper lobes of the lung. In four-legged animals, such as rabbits and cows, it's the dorsal portion of the lungs. Back in 1936, a couple of non-animal-loving scientists kept rabbits in an upright position (like a person) for 11 hours a day and demonstrated that their TB infections formed in a similar part of the lung to human infections. Not sure you'd get permission for that experiment these days.

Animals kept in low-oxygen environments have a slower disease progression, and the same appears to hold for humans. A study looking at TB in high-altitude villages in Peru suggested that transmission and disease rates decrease the further you are from sea level. So *M. tuberculosis* clearly likes oxygen. In fact, before antibiotics came along, one of the key treatments for TB was to artificially collapse a patient's lung. The idea here was that the lungs would get a nice rest and have a chance to recover from their infection. It was likely successful because it effectively suffocated the bacteria by depriving them of oxygen. Dr Harry Wilmer of *Huber the Tuber* fame went through this procedure himself a year before writing the book. At the time, he was an intern at a hospital in Panama, but was forced to return to the US to spend the following 11 months recovering at the Glen Lake Sanatorium. He includes collapse therapy in his book, describing it as a 'terrific wind' that accompanied the walls and ceiling closing in on Huber, bringing him within moments of suffocation. In the story, Huber survives, but presumably Wilmer's own personal Hubers all perished. If you're wondering, Wilmer went on to live to 88 after a long career as a psychiatrist in which he pioneered the use of group therapy to treat post-traumatic stress disorder.

I read a rather gruesome paper from 1936 that described a number of the available techniques for collapse therapy. The basic method involved anaesthetising the chest and inserting a giant needle into the pleural cavity. Air was then pumped in, squashing the lung and forcing it to collapse. If this wasn't possible, doctors could try paralysing the phrenic nerve or cutting a person's chest open (removing part of one rib in the process) to externally compress the lungs. And then there is plombage, in which a paraffin mixture was used to block off part of the lung. Eek. There's a lovely passage in the paper describing the results of complete thoracoplasty, which involves separation of the ribs from the chest wall to permanently collapse a lung:

A large proportion of the patients are rehabilitated to the extent that they are able to return to their former stations in life and to resume their former occupations or activities. Many women marry, and there is record of twenty-two who have borne children since their operation.

It's worth noting that the 'large proportion' mentioned here does not include the 34 per cent of patients who were dead by the time the report was written.

The Wellcome Library has among its collection of medical history resources a film from around 1925. It's a black-and-white silent recording from the Chicago Municipal Tuberculosis Sanatorium showing collapse therapy in all its barbaric glory. My favourite part, though, is the strange catwalk of sorts performed at the end to demonstrate the patients' recovery. The young women, dressed in white sheets to preserve their modesty and with their short hair gelled into fashionable waves, smile shyly and laugh with the doctors. Some faint scars and a slight wonkiness in their postures are the only signs of their operations. Their happy smiling faces suggest that they are all rather pleased with the results and have bonded with one another over their shared experiences. I suppose it was a form of group therapy, helping the patients to cope with any sense of isolation that came from dealing with a potentially fatal disease.

The idea that TB patients might help each other walk the arduous road to recovery isn't a new one. But with the wonder of modern technology, it's easier than ever for those affected by the disease to connect with each other. Charities such as Doctors Without Borders and TB Alert publish stories of TB treatment and recovery on their websites, normalising what must be a long and lonely experience for those affected by the disease. A web search revealed that there are a number of online forums available for TB sufferers. What amazes me reading the posts is how many people need to rely on strangers for information on their disease or the treatment they're receiving. Do doctors not speak to their patients?! Reading patient accounts like these made me realise how much more there is to treating TB than prescribing the right drugs. It's a team effort that involves not just the patient and his or her doctors, but support workers who ensure compliance with the long courses of medication, social workers to help address the vulnerabilities that put a person at risk of TB, family and friends, and fellow TB sufferers. Everything is better when we all work together. Same goes for scientific research.

One granuloma paper I want to mention is the result of a collaboration between 10 research institutions from the US and South Korea, involving 17 scientists. Heading up the project was Clifton E. Barry, III – an author on something like 200 papers and the holder of the most majestic name in TB research. When I try to think of who the most influential scientists in TB are, I come up with him first. Part of the reason for this is that whenever I've seen him at TB conferences, he's always been lounging right at the back dressed in his signature head-to-toe black like a scientific Milk Tray Man. He stands out in a sea of lab-casual. But he's also been responsible for some of the biggest papers in the field. In 2009, Barry's paper addressed a 50-year-old-plus question – are TB granulomas really hypoxic? After all, it's a theory underpinning much of the work that goes on into how *M. tuberculosis* survives during a latent infection, my own research included. Using a probe that stains hypoxic tissue and an oxygen electrode, the team was able to demonstrate that rabbit, guinea pig and macaque granulomas are all low on oxygen. In fact, even small, 3mm ($^{1/8}$in) rabbit lesions were severely hypoxic, with an oxygen concentration of 1.6mmHg (it's 100mmHg in the alveoli and 20-40mmHg in the lung capillaries). A more recent study from a London-hospital–based group led by Jon Friedland used PET (positron emission tomography)-CT scans on human TB patients to demonstrate that human granulomas are also hypoxic. They used a hypoxia tracer called [18^F]FMISO that has previously been used in tumour biology research to label regions with low oxygen concentrations, and found that many TB lesions label very strongly. The group linked the presence of hypoxic regions to the *M. tuberculosis*-dependent induction of a human protein called MMP-1. This protein is a member of the matrix metalloprotease family implicated as key drivers of the tissue destruction characteristic of TB and vital for transmission. So human TB lesions really are hypoxic, and this is linked to an up-regulation of the inflammatory signalling pathways that cause the lung to basically destroy itself.

This is clearly not great for patients, but the finding is exciting when you consider one of the most important laboratory models of what *M. tuberculosis* gets up to during a latent infection. If you grow a flask of *M. tuberculosis* in the lab and gradually expose it to a reduction in oxygen, it slows down and then shuts down. One way to investigate how a bacterium adapts to new conditions is to look at which genes are switched on and off under those conditions. Back when I was a student, microarrays were the new technology making all the promises when it came to gene expression. Later it was RNA-seq. Both techniques are doing roughly the same thing – detecting RNA and using it to determine which genes are switched on in a population of cells. I'm guessing that nearly everyone reading this book except for my mum and partner knows what RNA is, so skip over this little recap if you're not them. But just in case … So there's DNA – that's the genes, or the blueprint for everything that goes on in a cell. When the cell wants to switch on a gene, it copies the chunk of DNA that it needs into RNA and sends this messenger to the cell's manufacturing machinery to have it turned into proteins. The proteins then go off and do whatever task needs to be done. DNA is the blueprint; RNA is the messenger; proteins are the doers, makers, fixers. Techniques like microarrays and RNA-seq tell you what RNAs are present in a cell at a given time, which tells you what genes are switched on. </end science lesson>

When microarrays were first introduced, everyone was very excited. And then it became obvious that techniques like this generate tonnes of data that aren't always that useful. For starters, false positives are common. And then there's the issue that just because something is switched on doesn't mean it's vital for survival. It's like looking for the cause of insomnia by watching what people do when they're awake at night. Someone might sit up for a few hours trying to read *The Magic Mountain*, but that doesn't mean the book is responsible for the sleeplessness; it's merely a symptom. If you grow bacteria under hypoxic conditions, lots and lots of genes are switched on. But are these genes responsible for the bacterium's adaptation to hypoxia, or a by-product? Are they what sends the bacterium to sleep, or just the pillow it happens to rest its head on? My career has involved a lot of pillows. In my own experience, you can spend years studying one or two genes that you think might be important, only to find that the bacteria can cope pretty well without them after all. Surely they must do *something*, but it's trial and error trying to work out *what*. Basically, you can spend forever taking away a bacterium's pillows and discovering that it's still able to sleep just fine. Maybe its neck, gets a little sore, but you've not thought to check its neck so you don't realise. Or maybe it has prepared for this pillowless eventuality and simply pulls an almost identical one from a cupboard. At the same time, someone in another lab has decided to study the seemingly insignificant little square on the wall, only to discover that it's a light switch and the bacterium absolutely can't sleep if you take it away. They get a *Nature* paper, and you don't. Of course, you can always tell yourself that this particular light switch isn't all that interesting. Maybe the light switch is in a room that the bacterium doesn't sleep in except when it's being studied in a lab. Because – and here's the big problem with this entire field of research – no one has managed to unequivocally link the bacterium's adaptation to a low-oxygen culture flask to what happens during a human latent infection. Welcome to the wonderful world of scientific research.

As time goes on, it seems increasingly likely that the granuloma isn't a static structure encasing non-growing bacteria. There's actually evidence to suggest that it might not even be the home of bacteria during latent TB after all. I say 'after all' as if this is a recent discovery, but a paper from 1927 first raised questions about the location of *M. tuberculosis* during a latent infection. And then nearly everyone apparently forgot about it, which is a shame as the author sounds like he was right up there with the greats of the scientific world. According to his obituary, Eugene Opie was not the most physically robust of men. The author tells of a story from Opie's school days that I like, in which a student who would go on to become one of America's most noted TB specialists takes Opie under his wing. 'Brown was husky and strong and a leader in sports; Opie was seemingly frail and left out. Brown made the older and bigger boys include Opie in their groups and give him a chance in baseball and other sports.' Somewhat ironically, Brown would go on to die of TB at just 66 while Opie lived to 97.

It was during his time as a medical student at Johns Hopkins that Opie discovered the damaged region of the pancreas responsible for diabetes. This alone was enough to earn him a place in the scientific history books, but his research was to continue until close to his death. His obituary describes him as

> *a gentleman in every sense of the word. He was not expansive in social relations, but was always friendly and always helpful when his advice or assistance was sought. He had a quiet sense of humor that must have helped him many a time in his busy life. He could be firm in defending his opinions in controversy, but always temperately.*

As his career progressed, Opie's interest branched out from diabetes into, among other things, latent TB. Retirement couldn't stop him and he continued to work as a guest researcher until his final years. There's a quote by him that makes me think about my own short-lived research career: 'It is a fortunate circumstance that most of those who follow academic careers derive so much satisfaction from the doing of their work that they are unwilling to give it up.' It's Opie's research into what happens to *M. tuberculosis* during a latent infection that I'm interested in, though. In 1927, Opie co-authored a paper looking at the presence of live TB bacteria in a range of latent TB lesion types – 304 lesions, in fact, from 169 people. These lesions were in various states of 'healing', but the scientists were able to isolate living bacteria from them all. What was most interesting was that a quarter of the residual scars left over from a TB infection grew *M. tuberculosis*. This suggests that *M. tuberculosis* can persist in sites with no evidence of infection, and not just within granulomas.

The paper that first got me interested in Eugene Opie was from a Mexico-based research group led by Rogelio Hernández Pando. Incidentally, my research into Hernández Pando led me to a really interesting book on TB vaccine design that combines science with art (*The Art & Science of Tuberculosis Vaccine Development*, Oxford University Press). The foreword explains that 'the inspiration for the presented art images reflects the human impact of this devastating disease on the mainly socio-economically underprivileged individuals and populations of the world'. It's worth a look. Hernández Pando's 2012 paper is also a very creative and clever piece of work. His team used multiple techniques to investigate the location and viability of *M. tuberculosis* in individuals who died from reasons unrelated to TB and who had no clinical history of the disease. Mexico is a TB-endemic region, so it was not unexpected that 70 per cent of the subjects were found to have latent disease. What was really interesting was that there was viable *M. tuberculosis* present in the spleens, kidneys and/or livers of the infected people, in the absence of any inflammation or granuloma formation. Clearly there is life outside the granuloma.

One of the problems in studying what *M. tuberculosis* gets up to during a latent infection is that everything is happening inside the lungs of, hopefully, still living people. In the past, back when Huber was causing trouble, the only way to visualise TB was through an X-ray. TB would show up as a spot or shadow, which could be very difficult to differentiate from cancer and other diseases. Newer techniques combined with the use of animal models can solve a few of the problems in studying the granuloma. JoAnne Flynn of the University of Pittsburgh has been using modern imaging techniques to access the success of drug treatment in a cynomolgus macaque monkey model of TB. Remember the monkeys from the chapter on vaccine development? All of that vaccine work was built on earlier studies visualising TB granulomas. JoAnne's team combine CT scans with serial 2-deoxy-2-[18F]-fluoro-D-glucose (FDG) PET imaging. The CT scan provides the structural information about the lung while the PET imaging is used to identify regions that are actively doing something – like a heat-seeking camera, only you're looking for metabolic activity. FDG is a labelled version of glucose and is readily taken up by highly active cells – cancer or inflammation cells, for example. The CT scan spatial map of the lung can be overlaid with the granuloma hot spots identified by the PET imaging to track individual granulomas over time. The original idea was that JoAnne and colleagues could use this method to get a better understanding of how drugs affect different lesion types. What they discovered, though, ended up contributing to a new way of thinking about latent tuberculosis. By scanning the animals over time without drug treatment, they observed that the individual lesions were dynamic and independent of each other. In a single animal, some granulomas were stable while others grew or waned in size. It wasn't simply a case of the granulomas in a monkey with active disease gradually increasing in size in a uniform fashion. Even in the absence of drug treatment, the granulomas had independent trajectories.

'I don't even like to think about latent and active TB any more,' JoAnne says in answer to the very first question I ask. For a moment, I panic that I've committed some faux pas. Is she so bored of journalists asking about latent TB that she refuses to talk about it? Only then, she laughs and explains about the idea of a spectrum of disease. Latent and active TB are not separate entities divided by a 'go to sleep' or 'wake up' switch. Instead, latent TB is simply at the asymptomatic end of the spectrum. Its granulomas aren't necessarily any different from those involved in an active infection, although they're probably smaller and quieter on average. 'I think about the whole infection being a spectrum and clinically just being defined by the symptoms that you have, basically, which is kind of a sad way to diagnose anything, and then whether or not you can culture any bacteria.'

JoAnne was an author on a *Science Translational Medicine* paper that used PET/CT imaging to look at drug responses in macaques and humans. I'm going to be looking at TB drugs in more detail in a few chapters' time, but I'll mention this paper now anyway because I really like it. The study focused around some patients enrolled in a clinical trial looking at new treatments for drug-resistant infections pretty much untreatable with existing drugs. One patient had a 10-year history of the disease, and his infecting strain was resistant to every available treatment. He was scanned twice, two months apart, revealing that he had resolved one of his lesions while another had appeared in a new location. This waxing and waning of lesions was typical of all the patients. 'This backdrop of relatively constant overall disease in the lung highlights the dynamic and local nature of TB lesions,' the authors say. Within the same human, or the same monkey, each granuloma is doing something different.

The paper's images show a futuristic rendering of a person's chest and lung structures in blue against a black background. Have you seen the film *Weird Science*, in which two nerdy teenagers magic-up Kelly Le Brock using a home computer system and a freak bolt of lightning? I watched it again fairly recently and found it quite disturbing, but that's not my point. The computer program they use to design her reminds me of Flynn's CT renderings. Superimposed on Kelly Le Brock are what look like burning balls of fire. The heart shows up the brightest but then distributed throughout the lungs there are other orange-red blobs showing the extent of the patient's infection. It's surprising to see how different the lesions look between scans only two months apart. TB is far more heterogeneous than first believed. Even in latent disease, a person's granulomas are dynamic and constantly changing: a cycle of waxing and waning that ultimately favours the host but that can tip over into active disease should the immune system lose control. JoAnne told me how one granuloma is all it takes for latent TB to reactivate. 'How does the host control TB in 10 sites but not one site?' she says. 'I don't know, but that's the fun part. It's not like the other sites are sterile; they just don't reactivate. It's like they're their own little communities. Each granuloma is its own town in a state and they don't all do the same thing. To me, that's the fascinating part.'

Back when I started out in the TB research field, TB was talked about in binary terms. Latent TB versus active disease. These days, everyone accepts that it's in fact a spectrum ranging from individuals who have completely cleared the infection, to those with asymptomatic granulomas, to those who flit between active and latent disease, to those with fully fledged pulmonary TB. Lesions range from sterile tissue, to the hypoxic lesions I talked about earlier, to liquefied cavities containing huge numbers of bacteria, with everything in between. So I want to get away from the idea of Huber the Tuber sleeping away the years in his calcified home. Some cells may stop growing during a latent infection, and this is where models such as hypoxic dormancy come in (although hypoxia is only one of many ways to shove a cell into a non-growing state). But others can be happily dividing away, perhaps going through periods where they gain the upper hand over the immune system, and other times losing ground. The overall rate of division is matched by the rate at which the immune system can kill the bacteria, and an uneasy equilibrium of sorts is reached. In some people, latent TB might be a stable equilibrium in which Huber the Tuber really does spend his time reading *The Magic Mountain* and doing little else. In some, though, it would take very little to upset the balance in the infection's favour, and these are the people whom we need to keep an eye on. 'Within what we now consider to be asymptomatic TB, you know most people are going to be fine for the rest of their life,' JoAnne explains. 'The key question is how to identify those people who are not going to be fine. Who, in other words, are at risk of reactivation TB. And that remains a major question.'

With 2 billion latently infected people in the world, we can't go round imaging them all to look for those with poorly controlled granulomas. What we need instead is a simple test that, perhaps using either blood or urine, can determine who is controlling their infection and who is at risk of reactivation. This is where biomarkers come in. A biomarker is a bit like a fingerprint left behind by a pathogen – or any other disease, for that matter. With TB, biomarkers can be bacterial in origin: parts of the cell wall, mycobacterial DNA or protein antigens, for example. Or they can be something on the host's side, such as certain immune genes being turned on or a specific pattern of immune proteins in the blood. Over the last decade, there have been a few dozen studies using blood to look at what genes the host switches on in response to a TB infection. One interesting paper came out in *Nature* in 2010, from Anne O'Garra's London-based research group. Anne's group generated profiles of gene expression from the blood of active TB patients, latent TB patients and healthy controls. A signature based on the expression patterns of 393 genes could reliably detect active TB patients from an independent cohort of patients. Interestingly, a number of latent TB patients – between 10 and 20 per cent – had signatures close to that of active TB. The idea of TB being a spectrum of disease can explain this observation.

In a recent paper from Robert Wilkinson's group, JoAnne's macaque model was used to demonstrate that, of 35 asymptomatic HIV-TB co-infected patients, 10 in fact had lung abnormalities suggestive of subclinical active disease. Their infections were well on the way to reactivation, even if their symptoms hadn't caught up with them. Where possible, latent disease in HIV-positive individuals is treated with isoniazid as a way to prevent the high rate of reactivation among this population. However, a single drug is unlikely to be successful in patients with subclinical active disease – they would do better with the standard four-drug treatment. Treatment of latent TB is a controversial area. If you're diagnosed with it in the US, for example, it's routine to treat with isoniazid, and this is likely one of the reasons why TB rates in the country continue to decline. But in high-burden areas, the importance of transmission may outweigh the benefits of treating latent TB – what's the point in clearing someone's latent infection if the majority of cases are due to new infections anyway? Other critics suggest that treatment with a single antibiotic will lead to the selection of isoniazid-resistant mutants. One 2013 paper showed that there's a small window of opportunity in which isoniazid preventative treatment can yield a beneficial result. But, even when rolled out optimally, it will eventually drive up drug resistance. On top of that, you have some nasty side effects, which means that taking isoniazid unnecessarily is not anyone's idea of fun. I'm fairly certain that, if I were diagnosed with latent TB, I would refuse treatment unless I had some co-infection or morbidity that made it likely I would reactivate. Or if there was a simple test to reveal that reactivation was on the cards, of course. While we're not there yet, it's an area to keep an eye on in the future.

Huber the Tuber ends with an illustration of Huber contentedly sleeping in his calcified granuloma while a member of the immune system barricades him inside. A prisoner of war who will hopefully remain in jail indefinitely. 'But Huber is still alive. Maybe a few scattered Tubers of the virulent strain escaped to some other fortified airsacs. They will live imprisoned in calcium because the Home Guard will prevent their escape with barbed wire.' Not every mycobacterial cell present in a latent infection is like Huber – sleeping the years away and waiting for his world to change. Some of them are, while others are doing something else entirely. So coming up with new ways to treat TB, wherever it is on the spectrum, is unlikely to have a one-shot solution. Each population of cells will have different weaknesses ready to be exploited by us humans. To know what those weaknesses are, we first need to understand how exactly *M. tuberculosis* survives in the face of challenge from the host's immune system. How does Huber the Tuber hide from the cells that would recognise him and raise the alarm? What defences does he employ to avoid being killed by the Home Guard, and what does he eat as he settles down with that copy of *The Magic Mountain*? Answer these questions and maybe we'll spot an Achilles heel that can be exploited by new drugs and treatment strategies.

CHAPTER EIGHT

Growing Fat on the Atkins Diet

If *M. tuberculosis* were a person, I imagine that its diet would consist of long periods of bin-diving for half-empty vitamin bottles and the occasional haggis binge. For any US readers who aren't au fait with the culinary horror that is haggis, this is probably because your country banned the import of authentic Scottish haggis as it contains lung meat. Also, sheep intestines, liver, heart and some oats for good measure. Back in 1988, the *British Medical Journal* featured a paper looking at whether this national dish was contributing to the rather high incidence of coronary heart disease in Scotland at the time. Haggis, you see, has quite a lot of cholesterol in it. One small serving contains around 150 per cent of a person's recommended daily cholesterol intake. Serve up some haggis, and *M. tuberculosis* would be in gastronomical heaven. I imagine that, were *M. tuberculosis* a person, it would have gallstones and be carrying quite a bit of excess padding.

Even as a bacterium, *M. tuberculosis* is on the chubby side. Mycobacteria are unique among bacteria in that they're surrounded by an unusually thick, lipid-rich cell wall. This cell wall is a waxy, almost impenetrable barrier. In fact, I spent a large proportion of 2008 trying to find ways to disinfect *M. tuberculosis* cultures in the laboratory so that we could safely dispose of them without killing anyone. Our tried-and-trusted disinfectant had been banned due to being horribly toxic to all known life forms, *M. tuberculosis* and humans included, and disinfectants that worked on other bacteria weren't proving to be even half as effective. All thanks to that cell wall. If *M. tuberculosis* was a person, its outer coating would be around 6cm ($2^{1/2}$in) thick. Imagine coating your entire body in a 6cm-thick layer of solid butter and liquid margarine, imbedded with lumps of protein and sugary cake sprinkles. Congratulations, you're now looking not very much like a mycobacterial cell but you're quite delicious. I'm guessing you're also quite resistant to most antibiotics, just like *M. tuberculosis*. For an antibiotic to work, it has to be able to get inside.

It's not just a barrier, though. The mycobacterial cell wall is a complex, dynamic structure that scientists have dedicated entire careers to understanding. It protects the cell both passively, by keeping toxic substances on the outside, and actively, by interacting with the host immune cell to reprogramme it in the bacterium's favour. And when it's done ensuring the bacterium doesn't die, the cell wall is also going to help *M. tuberculosis* ensure it can find enough food, stolen straight from the macrophage's own supplies. If *M. tuberculosis* were a person, it would be a pro at the buffet table. I imagine he'd be the sort to position himself perfectly in preparation for the arrival of the new platters of food; someone who can instantly tell the deep-fried mozzarella balls from the battered veggie bites, and spot the bacon-wrapped cocktail sausages hiding behind the egg sandwiches. And then, in a somewhat surprising development, person-*M. tuberculosis* would go on to fashion himself a nice cloak from stilton and macaroni cheese, wearing it proudly as he continued to feast. He certainly is an odd character. But there's no denying that M. tuberculosis is the master of the quick costume change and is always dressed for dinner. Finding itself inside a host macrophage, the bacterium rapidly induces a range of genes that remodel it in preparation for life in a cell dedicated to killing. Part of this response alters the composition of the cell wall. It's all part of a realignment of a whole range of cellular structures and processes – in both the bacterium and host. What *M. tuberculosis* eats impacts on more than just its waistline.

In the laboratory, *M. tuberculosis* is cultured in what is basically a flask of water in which various nutrients have been dissolved, the most abundant of which is the bacterium's source of carbon. A couple of years ago, there was a man in the news who'd invented a drink he claimed was an all-in-one nutrition shake. Soylent promises to free your body of the need to eat with its neutral-tasting concoction of soy protein, algal oil, isomaltulose, vitamins and minerals, and people[*] (not really). For around $14 a day, you could theoretically live entirely on Soylent shakes, but would you want to? *M. tuberculosis* isn't quite so fussy, what with being a single-celled organism. So it happily eats its Soylent equivalent and grows and grows until it runs out of food or space. Most commonly, glycerol is used as a carbon source in the lab. You might have come across glycerol through its use as a sweetener in the food industry. *M. tuberculosis* has a real taste for glycerol and, given a constant supply, grows in size and doubles in number at the blistering speed of one division every 20 hours or so. Which actually isn't very blistering at all. The doubling time of most bacteria is measured in minutes. *Escherichia coli*, for example, doubles every 20 minutes. The bacterium behind strep throat is almost as fast, at around 40 minutes. So *M. tuberculosis* is the garden snail of the microbial world.

It's not alone. The *Mycobacterium* genus includes a number of other species capable of causing disease in humans. What's interesting here is that these disease-causing microbes also hang out among the slow-growing mycobacteria, suggesting a link between slow growth and pathogenicity. Does slow growth make it easier to survive inside a host cell or does surviving inside a host cell push a pathogen towards slower growth? (What came first, the chicken or the cholesterol-laden egg?) I asked scientist Dany Beste about the slow growth thing. She works on mycobacterial metabolism, which basically makes her a bacterial dietician, although I didn't say this to her face. 'As for why [*M. tuberculosis*] grows slow, I feel that's the million-dollar question,' she explained. 'I've always thought it's because [*M. tuberculosis*] is so fat! Most bacteria contain a fraction of the lipid content that *M. tuberculosis* contains. Generally, cells with high lipid content are slow-growing.' One possibility is that, in cutting itself off with that

---

[*] *Soylent Green* is a 1973 science-fiction film in which a dystopian population survives on rations made from human remains.

thick wall, *M. tuberculosis* has also made it difficult for nutrients to get inside. Putting bouncers on the door and maintaining a selective guest list can seriously slow things down. But I personally find it hard to imagine that, during *M. tuberculosis*'s long history of co-evolution with humankind, it would have put up with something as fundamental as its exceedingly slow growth rate if that didn't provide it with some kind of benefit. I suspect the slow growth is part of the various adjustments made by *M. tuberculosis* to the host environment. A kind of 'slow and steady wins the race' when you're a chronic pathogen that not only needs to avoid being killed by the immune system but for much of its lifecycle survives in an environment low on nutrients.

Even at the very beginning, when the macrophage first engulfs *M. tuberculosis*, that thick and waxy bacterial cell wall is playing a key role. There are numerous routes mycobacteria can use to gain entry into the host cell, all of them involving interactions between the bacterium and host cell receptors. Off the top of my head, I can think of a dozen films where the characters deliberately allow themselves to be captured in order to get inside the enemy base. This is kind of what *M. tuberculosis* does. If it's going to be ingested by the immune system, it wants to do it on its own terms. Then, once it's on the inside, the bacterium can attempt to turn things round to its own advantage and not die. Depending on the type of cell that's eaten it and the route it has taken on the way in, this may or may not be successful. One port of entry might toss the microbe into a compartment where it will rapidly be digested, but another might allow it to circumvent macrophage activation.

Inside the host cell, *M. tuberculosis* finds itself enclosed in a little membrane-surrounded bubble called the phagosome. It's an organelle designed to kill that comes kitted out with an array of weaponry. One of the earliest bactericidal activities of the phagosome is its acidification to below pH5. On top of the low pH, the phagosome also fills up with hydrolytic enzymes that can digest foreign invaders. Then the macrophage prepares to activate its primary weapon – the superoxide burst. This is a piece of machinery transported into the membrane of the phagosome to generate toxic reactive oxygen species ('free radicals' in skincare advert-speak). *M. tuberculosis* wants to avoid all these killing mechanisms. So it tricks the macrophage into arresting the development of the phagosome at an early, non-lethal stage. This, in part, is triggered by the actions of cell wall lipids. In 2004, David Russell's research group in upstate New York attempted to identify *M. tuberculosis* genes involved in this arrest of phagosome maturation. Using a library of *M. tuberculosis* mutants, they infected human macrophages and used microscopy to identify those that ended up in fully formed, bacteria-killing phagosomes. These were the mutants that had lost the ability to prevent the phagosome from fully maturing. Identify the mutated gene and you identify one of the bacterium's mechanisms of survival. They discovered genes with a range of functions, from transporters to lipid biosynthesis genes. Some were involved in making phthiocerol dimycocerosates – the most abundant lipids in the mycobacterial membrane.

M. tuberculosis surface lipids are a complex affair. Diagrams of the mycobacterial cell wall remind me of something growing at the bottom of the ocean, tethered to a deep-sea vent or coral reef. It certainly looks very pretty, with the branched mycolic acids and arabinogalactan rendered in bright colours. Depending on the scientist's flair for art (or their lab's media budget), the diagrams might contain colourful spirals all interconnecting with the zigzagging lipids and layers of peptidoglycan. Or it might just be some circles and squares with wonky lines drawn between them. Phthiocerol dimycocerosate (PDIM) is usually drawn as complicated hairclip-like structures that appear to have been stabbed into the mycobacterial envelope from the outside. All right, so it's not the most visual of subject matters, but hopefully you get the rough idea. In any case, PDIM is interesting as it's made by all the members of the MTBC but isn't found in non-pathogenic mycobacteria. PDIM is also a giant pain in the backside when it comes to the scientific study of *M. tuberculosis*. Imagine you spend a year or more making and characterising a gene deletion in *M. tuberculosis*. And you discover that knocking out this specific gene leads to a severe attenuation in the bacterium's ability to cause infection. Brilliant, you've discovered something really important! Only, when you try to add the deleted gene back in – a process known as complementation – you can't restore the original level of virulence. What has happened is that during all the months you spent making your mutant, the strain has sneakily gone and lost the ability to make PDIM. All the results you've carefully gathered together have nothing to do with the gene you deleted. It was PDIM all along.

The reason for this is that making a great big lipid tree takes a lot of energy – wasted energy, if the bacteria are growing in the lab and not a living host. In the culture flask, a strain that has stopped making PDIM will quickly outcompete the original strain, as it is able to use its resources to grow slightly faster. PDIM-deficient variants are the Japanese knotweed of the TB research world. They're also quite likely responsible for invalidating the conclusions of any number of scientific papers published despite a lack of successful complementation. Worrying stuff. So, we know PDIM has an important role in virulence, but what exactly is it doing? In 2009, Christophe Guilhot's lab demonstrated that PDIM is involved in the initial invasion of macrophages. In their model, PDIM is inserted into the plasma membrane of the host cell, where it modifies the membrane's biophysical properties. This encourages/tricks the macrophage into engulfing the bacterial cell and later plays a part in preventing phagosome acidification. More recently, David Sherman's lab looked at the role of PDIM post-internalisation, during the pivotal first few weeks of an infection before the adaptive immune system comes into play. They used a replication clock to calculate bacterial growth and death rates during a mouse infection. A replication clock is an unstable ring of DNA (plasmid) that is lost at a steady rate as the cells divide. By looking at how much of the plasmid remains, it's possible to calculate how many times a cell has divided. The scientists discovered that there is almost no death of *M. tuberculosis* in the first two weeks of infection. During this time, the bacteria survive in the permissive macrophages, arresting phagosome maturation to limit exposure to acidic and toxic conditions. It's the glory days of an infection, at least from the perspective of *M. tuberculosis*. Unless, of course, it is unable to make PDIM. When Sherman's group used their replication clock to look at the death of a PDIM-deficient strain, they discovered that it is rapidly killed during the initial two-week period. Another study showed that PDIM-deficient bacteria could be found in acidified phagosomes, indicating that PDIM contributes to the remodelling of the phagosome to turn it into a nicer place to live. How it performs this role isn't fully understood, but it could have something to do with changing the properties of the host phagosome membrane to prevent the acidification machinery from working.

A slightly different role for PDIM was proposed by Lalita Ramakrishnan's lab and published in *Nature* in 2013. The theory here is that *M. tuberculosis* uses PDIM as a bacterial equivalent of a balaclava, using these abundant lipids to mask features that would alert the wrong kind of immune cell. Without PDIM, the bacteria are naked and instantly recognisable. The alarm is tripped early in an infection, while the bacteria are still in the upper parts of the lung. Once the immune system reaches DEFCON 1, it becomes much harder for *M. tuberculosis* to make itself comfortable inside the macrophage. The immune cells are at too high a level of alert and start to shoot on sight. The upper part of the lung is permanently at DEFCON 2, so it doesn't take much to push the immune cells here over the edge. So *M. tuberculosis* uses PDIM to mask its identity for long enough to creep into the deeper regions of the lung. It's a bit like Norfolk down there, or New Jersey. Whereas the upper respiratory tract is full of microbes, constantly stimulating the macrophages and maintaining them in a state of high alert, the lower respiratory tract is relatively sterile. The police force here sits around chatting and drinking coffee. So it's easy enough for *M. tuberculosis* to find itself some nice, permissive macrophages that it can make its own. David Sherman told me how PDIM not only has multiple roles that likely vary between different stages of an infection, but that we've only scratched the surface when it comes to understanding this molecule.

Even with PDIM, the bacteria can't keep creeping beneath the radar forever, though. Eventually the immune system is going to catch on and start activating the macrophages. Then things get a whole lot harder for *M. tuberculosis*. Activation of a host macrophage leads to alternations in the phagosome, including the generation of reactive nitrogen intermediates – similar to the reactive oxygen intermediates above, only meaner. Thankfully for *M. tuberculosis*, the activation state of the macrophages exists as a spectrum, and while *M. tuberculosis* is killed by some cells, others just become less comfortable. One of the roles of macrophage activation is to restrict an invading bacterium's access to food. In a permissive macrophage, *M. tuberculosis* is able to access the macrophage's own nutrient sources. But activation means that larder privileges are revoked. This is where the adaptability of its metabolic pathways comes in handy. One of *M. tuberculosis*'s strengths has to be wringing everything it can out of its environment and ensuring that it uses all the nutrients available in the most efficient manner.

Growth of the bacteria in the inhospitable lung environment is nothing like the predictable, controlled growth that we see in the laboratory on our highly defined medium. And *M. tuberculosis* is certainly not using glycerol. As far back as the 1940s, scientists were wondering if the way they were growing *M. tuberculosis* in the lab was in any way representative of what happens in the host. The first evidence that *M. tuberculosis* uses a rather unusual source of carbon in the host came from a 1955 paper by William Segal and Hubert Bloch. The scientists isolated bacteria from the infected tissues of mice and found that they were very different from a lab-grown culture. To get an idea of what food source the isolated bacteria had been using in the host, they offered them a range of different options and looked at how keenly they responded. Like a fingerprint of what the bacteria were doing in the host, their response to the different food sources revealed what they were used to metabolising. When it came to glycerol, glucose and a range of other substrates, the bacteria showed no interest. The authors described the cells as sluggish and non-reactive. Alone, these results give the impression that the bacteria were existing in a reduced state of activity, with all their internal reactions ticking over rather than revving at full speed. They just didn't seem that hungry. Only when presented with fatty acids (the component parts of lipids), the bacteria gobbled them right up. The pathways required to use lipids were all warmed up and ready to go. Inside the mice, *M. tuberculosis* was partaking in a microbial version of the Atkins Diet.

When the *M. tuberculosis* genome sequence was published in 1998, it was no surprise that it encoded examples of lipid biosynthetic machinery. What was surprising was that there were examples of every known lipid biosynthetic system, including enzymes usually found in plants and mammals. The bacterium also makes around 250 enzymes involved in fatty acid metabolism, compared to just 50 found in *E. coli* bacteria. In some cases, it contains more than one enzyme to do the same job. Bacteria don't hang on to backup copies unless the function they are performing is vital for survival. The importance of these pathways is underscored by studies showing that deleting some of their number results in catastrophic defects in the bacterium's ability to cause infection. One such example is isocitrate lyase (ICL). *M. tuberculosis* contains two copies of the *icl* gene, both of which can be deleted from the genome with little effect. Just so long as the mutant strain is grown on glycerol. Give the mutant fatty acids, however, and the loss of the *icl*s is incompatible with life. In a mouse model of infection, the mutant is rapidly cleared. It isn't just a bit rubbish at causing infection, it is incapable of surviving in a living host. This makes it one of the most severely attenuating mutations in a mouse model, underscoring the fundamental role of fatty acid metabolism during an infection. Because of the importance of ICL, it has been proposed as a potential drug target. It's particularly attractive because this enzyme is absent in mammalian cells, limiting the possibility that a drug targeted to ICL would also prove toxic to the human patients.

Further supporting ICL as a drug target are more recent studies suggesting that its role goes much further than fatty acid metabolism. In 2013, a team from the Weill Cornell Medical College, New York, were looking at the response of *M. tuberculosis* to antibiotics. They used a technique called metabolomics, which is the equivalent of examining the crumbs under the buffet table to get an idea of what people have been eating. Metabolomics lets you see what metabolic pathways are switched on at a given time by detecting the chemical fingerprints that they leave behind. The group led by Kyu Rhee wanted to know how antibiotics affect the bacterium's metabolism, so they exposed the cultures to three different TB drugs. Despite these drugs targeting very different functions (RNA synthesis, protein synthesis and the cell wall, respectively), all three induced similar remodelling of the organisms' metabolic pathways, including an induction of the bacterium's ICLs. The double *icl* mutant, it turns out, is also far easier to kill using antibiotics.

The way in which antibiotics kill bacteria is two-fold. First, there are the vital processes that the antibiotic specifically attacks. These are the main targets in the drug's crosshairs. But drugs can also kill in a sneakier manner by forcing bacteria to produce their own downfall in the form of free radicals. These free radicals are products of the bacterium's own metabolic machinery, produced as its pathways are knocked out of sync by the antibiotic. I imagine them like a deranged game of Pong, in which every bounce doubles the number of balls and each ball is a miniature bomb capable of destroying anything it hits. When it's working properly, a bacterial cell keeps everything under control. Add an antibiotic, though, and it loses its grasp on the free radicals. They bounce all over the place and attack DNA, proteins and lipids, damaging them beyond the cell's capacity for repair or replacement. Kyu Rhee's theory is that the ICL enzymes are playing a vital role in preventing the build-up of free radicals. By regulating the cell's metabolic pathways, it's possible to balance out free radical production and minimise the damage they cause. This is a new role for the ICLs, as well as being a really interesting link between the cell's metabolism and antioxidant defence systems.

Another vulnerable point in *M. tuberculosis*'s metabolic pathways could be its surprising ability to use cholesterol as a carbon source. Similar to the *icl* double mutant, knocking out enzymes involved in the uptake or metabolism of cholesterol leaves the bacterium unable to survive during the later stages of a chronic infection. But ... *cholesterol*? No other pathogens are known to use cholesterol as a food source; it's unheard of. Yet *M. tuberculosis* clearly does. Its unique taste for cholesterol is, in fact, likely to be a major strategy that allows it to survive inside the macrophage. Early on in the infection, the bacterium has lots of choice when it comes to what it can eat. But once the macrophages become activated, it needs to find an alternative. That alternative is cholesterol stolen directly from the host cell itself. Then it's just a case of clinging on the best it can until the immune system loses its grip on the infection. Things get easier for *M. tuberculosis* gastronomically speaking as the infection progresses. Once the granuloma starts to break down, it forms a mushy substance at its centre known as 'caseum' – from the Latin word for cheese. Yum, yum, YUM. It's a feast rich in lipids that allows the bacteria to stop scrabbling around for every last scrap of food and get on with the business of rapidly dividing (as rapidly as something as slow as *M. tuberculosis* can go, at least). Eventually, the whole granuloma liquefies and ruptures, spewing out millions of fat, happy bacteria that can be coughed up and spread to new hosts.

I read a really interesting paper from 2010 that took advantage of the last-ditch treatment for extensive TB – excision of the infected regions of the lung. While the patients recovered from their surgery, the scientists dissected the granulomas out of their freshly removed lung tissue and extracted the hosts' RNA. Using microarrays, they were able to get a snapshot of what human genes were switched on in a caseous, soon-to-liquefy granuloma. They discovered that a large number of proteins involved in host lipid metabolism are upregulated – that's *host* proteins, not bacterial proteins. They hypothesised that *M. tuberculosis* is somehow realigning the host's lipid metabolism to provide itself with an unlimited source of food. The theory is that components of *M. tuberculosis* trick the host cells into sequestering lipids in the form of lipid droplets. 'Foamy macrophages' rich in these droplets accumulate at the centre of the granuloma. Later, the foamy macrophages die. Lots of cholesterol is released and builds up as caseum, which the bacteria can use as a source of energy and building materials to maintain that important cell wall. It's a clever strategy, which makes *M. tuberculosis* stand out as special among pathogens. It could also be its Achilles heel. Cholesterol pathways are interesting drug targets, not just because cholesterol is an important nutrient but because using it requires extremely tight regulation. As cholesterol is broken down by the cell, its component parts have the potential to build up and intoxicate the bacteria. PDIM, incidentally, gets in on the act here too, by acting as a sink for unwanted bits of cholesterol. Interfere with cholesterol degradation pathways, and not only can you remove the bacterium's food of choice but you can also force the cell to poison itself. Cholesterol, it seems, can both save and doom *M. tuberculosis* depending on how it is used.

The same theme follows through into our own human relationship with cholesterol. One-third of UK and half of US adults have more cholesterol in their blood than is deemed 'healthy'. But cholesterol has more wide-ranging effects on our bodies than making our jeans and arteries a bit too tight. While cholesterol is needed for the correct function of immune cells, high levels are believed to lead to a skewing of the immune response and impaired immune cell function. So in our world where cholesterol is basically a swear word, what does hypercholesterolaemia mean for the immune system's ability to fight a TB infection? In 2008, Hardy Kornfeld led a team looking into immunity to TB in mice fed a high-cholesterol 'Western' diet (burgers and fries, perhaps?). In these mice, a TB infection proved more serious, with more lung damage and a higher mortality. The problem was that the mice were sluggish in mounting a T-cell response to the infection, meaning that the immune system was on the back foot right from the beginning. Instead of quickly switching on that strong cell-mediated immune response required to fight off the infection, the immune cells did a poor job of reacting correctly to *M. tuberculosis* antigens.

Recently, there's been talk that statins – commonly prescribed to lower 'bad cholesterol' in patients at risk of heart attack or stroke – might also protect a patient against active TB. Chien-Chang Lee of the National Taiwan University Hospital led a study retrospectively looking at 8,098 patients diagnosed with TB between 1999 and 2011. They found fewer cases of TB among patients treated with statins than in untreated patients. In fact, the risk of developing TB was almost halved in patients undergoing statin treatment. Of course, this is an association and not a proven link. Other differences between statin users and non-users could have been responsible for the different rates of TB, but it does raise some interesting questions. In 2014, a team led by Cape Town-based Frank Brombacher published a study suggesting more directly that statins can augment the host immune response to TB. They isolated macrophages from the blood of patients receiving statins for familial hypercholesterolaemia and infected them with TB. Macrophages from statin-treated individuals were better at surviving compared to those isolated from the control subjects. This wasn't a human trial, so isolated macrophages were as close as Brombacher could get to an actual person. Instead, the scientists tested the statins in *M. tuberculosis*-infected mice. The animals showed reduced levels of bacteria in their lungs and less pathology. Brombacher puts these results down to the statins giving immune cells a helping hand in triggering phagosome maturation and by encouraging autophagy, or the suicide of infected cells. More recently, a study led by Petros Karakousis of Johns Hopkins Medicine looked at TB-infected mice treated with standard TB drugs in the presence and absence of simvastatin, a statin used by millions of people. They found that the time taken for a mouse to be cured of TB was reduced from 4.5 to 3.5 months by statin treatment, and the relapse rates were significantly reduced following 2.5 and 3.5 months of treatment. Statins don't appear to have a direct effect on *M. tuberculosis*. Instead, they are what's known as a host-directed therapy – a therapy that helps the host fight the infection using its own tools. There's now interest in taking statins into human trials to test whether they're able to augment host protection against TB, with particular emphasis on reducing treatment duration. Of course, there's a lot of safety considerations standing in the way, in particular how statins can be safely combined with anti-TB drugs. But as medications go, statins are cheap and relatively safe, so in all, a promising start.

Cholesterol does have a bit of a bad reputation these days, though, and it's not entirely deserved. So in an effort to maintain a healthy balance in this chapter, I want to mention the problem of too little cholesterol. A couple of studies have shown an association between low cholesterol and active TB, or mortality among TB patients with disseminated disease. So a team of doctors in Mexico City decided to test whether supplementing TB patients with a cholesterol-rich diet could help them recover. Ten patients were selected, with a body weight range of 38.5 to 59.5kg (85 to 130lb) and cholesterol levels below the median value for Mexico as a whole. They were given a 2,500 calorie diet comprising 800mg of cholesterol per day, derived from foods such as butter, beef liver and egg yolk. Compared to a group of patients receiving the same number of calories but less cholesterol, the experimental group showed signs of being cured much faster. It's worth noting that while their levels of LDL cholesterol increased, they did not reach levels believed to put the patients at risk of cardiovascular disease. Instead, the cholesterol-rich diet may have replenished cell membrane cholesterol needed, for example, for the correct functioning of the immune system. With cholesterol constituting up to 30 per cent of a cell membrane, it has a big part to play in the structure and function of the cell. Too little cholesterol, and cell can't work like it's meant to.

It's well known that nutritional status in general can influence the ability of the immune system to do its thing and that poor nutrition has a knock-on effect on TB. While the African continent has HIV feeding its TB epidemic, India has endemic undernutrition. Chhattisgarh, in particular, has a huge problem with undernourishment. This rural state has a population of 20 million, a third of which are indigenous people suffering from high rates of poverty and marginalisation. A 2013 collaboration between the Himalayan Institute of Medical Sciences and McGill University, Canada, found that 67 per cent of male TB patients and 80 per cent of females in Chhattisgarh were undernourished (with a BMI below 17). Fifty-two per cent of patients had stunting as a result of chronic undernutrition. Undernutrition was highly associated with an increased risk of death and often remained a problem, even following successful TB treatment. Based on this, the authors of the paper called for nutritional support during TB treatment among this population. In 2015, Chhattisgarh became the first Indian state to set aside part of its budget for providing nutritional supplementation to TB patients. Other parts of India have attempted something similar, although ready-to-eat food packets failed to go down well in Mumbai, with half of MDR-TB patients refusing them on the basis that eating the same thing every day was monotonous. I've read case studies of individuals who have benefited from nutritional supplementation during TB treatment, but when it comes to clinical trials, there isn't actually much evidence for an improvement in outcome. I suppose it depends on how you measure improvement. Death rates among TB patients aren't the only statistics we need to look at. What about quality of life and protection against costs accumulated as a result of TB? On top of this, there's evidence from TB control programmes in Afghanistan that food assistance can help encourage adherence to treatment.

From India to China, we can find yet another example of how underlying patient health can impact upon TB. Here, it's diabetes that threatens TB control efforts through its deleterious effects on the immune system. While a small proportion of diabetes cases are down to the immune system attacking insulin-producing cells (type I diabetes), the big problem is the worldwide increase in type II diabetes resulting from the body losing the ability to use, and later to make, insulin. China has the world's largest type II diabetes epidemic, with more than 100 million people affected. Thanks to its population of over 1.3 billion, it also harbours more cases of TB than any other country excluding India. Every year there are around 1 million new cases of type II diabetes – 10 per cent of the world's total – and close to 40,000 deaths. Someone with diabetes is three times more likely to develop active TB, so China finds itself in a worrying position despite its success over the past two decades in reducing its incidence of TB. It's been estimated that by 2035 – the deadline for End TB's ambitious targets for reducing TB incidence by 90 per cent – the total number of people worldwide with type II diabetes will number 600 million.

With one parent already on medication for the condition, I'm well aware that I too may become part of this troubling health statistic. I don't know about you, but I don't much fancy becoming the subject of one of those news reports on the growing diabetes epidemic. You know, the ones where they show the back view of a couple of big-bottomed individuals in poorly fitting clothes, waddling away from the camera with their heads out of shot? I've heard the condition referred to as 'diabesity' due to its unholy partnership with obesity, but I personally find the term vaguely insulting even though it isn't aimed at me (yet). It seems to play into the stereotype that diabetes is a disease of lazy excess. Low- and middle-income countries have their killer infectious diseases; the Western world eats too many donuts. In reality, it's a complete myth that diabetes is a disease of rich nations with too much food. Of 382 million individuals with diabetes across the globe, 80 per cent of them live in low- or middle-income countries. In fact, the distribution closely overlaps with the incidence of TB. It's scary that, despite all of this, the relationship between diabetes and TB has often been overlooked even though a number of studies have shown a bidirectional link in which each condition makes the other worse. Should countries be screening patients with TB for diabetes to enable treatment of both diseases? Or screening patients with diabetes for latent TB and offering them preventative prophylaxis, much as is recommended for HIV-positive individuals? Quite possibly, yes, but there have yet to be any trials that can say for sure what works and what doesn't.

Diabetes-TB is yet another example of how TB is far more complicated outside of the laboratory. As a microbiologist, I spent my days investigating the bacterium under strictly defined and controlled conditions: one food source in excess, and certainly no other diseases or bacteria interfering with the results of my experiments. Only, real-world *M. tuberculosis* doesn't survive on glycerol like in a culture flask – it uses a whole mixture of nutrients scavenged from the host, and has evolved complex pathways to ensure that it doesn't go hungry. For a single-celled organism, it's a complicated beast, capable of adapting to changing conditions as an infection progresses. Its unique ability among pathogens to use cholesterol as a source of food is just one of the mechanisms it has to help it survive inside a cell that's evolved specifically to kill infectious agents. The complex, lipid-rich cell wall is another, with its constant communication with the host to ensure that the infection progresses according to the bacteria's plans. In turn, the success of the bacterial mechanisms is influenced by the overall health of the patient – cholesterol, nutritional status, host metabolic state. TB as a disease is so much more than a bunch of bacteria cocooned inside a lonely granuloma. Research into areas such as mycobacterial metabolism has the potential to highlight areas of weakness that could be exploited by new treatments; more often than not, it reveals another reason why this pathogen is so hard to kill.

CHAPTER NINE

Killing the Unkillable

The streets of London are eerily silent. Smashed cars with their doors and boots open make rusty traffic jams; weeds sprout from the cracked pavements. An escaped zoo lion tentatively pads across a scattering of broken glass and peers inside Cool Britannia at the shelves of London bus snow globes, Union Flag umbrellas and rubber ducks wearing bearskin hats. It's the only shop that's not been looted because, when the zombie apocalypse hits, the last thing anyone needs is a plastic policeman's helmet and a mug featuring a picture of Prince George. The sound of a bolt crunching open sends the lion loping back to its den. A door creaks ajar and a pale, vitamin D-deficient man blinks up at the grey sky. He's middle-aged with a scraggly beard and army fatigues stained with the tinned beef stew that he's lived off for the last six months. Let's call him Jeff. Jeff is one of the few survivors who are going to repopulate the decimated planet. Jeff is the future of humankind.

Among every population of bacteria, there are Jeffs. Individuals who, genetically speaking, are no different from all their friends and neighbours. Maybe if bacteria could speak, they'd call their Jeffs 'a bit strange' or they'd accuse them of wasting their life preparing for an apocalypse that was never coming. Hiding indoors, avoiding crowded areas, stocking up on meals in a tin and bottled water. Jeffs are the bet-hedgers, the survivalists, the preppers. But when the end of the world actually does arrive, Jeffs are the ones who don't die. In the microbial world, Jeffs are the 'persisters'. Persisters make up a small proportion of a bacterial population and, thanks to rearrangements of their cellular machinery, can survive high doses of antibiotics. But there's nothing in their DNA that makes them special. Once the danger has passed and they get back to multiplying, their progeny is still as susceptible to the antibiotic as the majority of the original population. They're not genetically resistant but their continued survival can give a bacterial infection the opportunity it needs to develop resistance. They're why it's so important that we take the full course of an antibiotic. Because if we don't make sure all the Jeffs are dead, they're just going to re-multiply, and next time round, they might come better prepared.

Persister cells are not unique to *M. tuberculosis* but a global phenomenon employed by all species of bacteria. Everyone, it seems – from humans right down to tiny microbes – knows someone like Jeff. The formidable penicillin expert Gladys Hobby was the first to identify these cells in 1942, but Joseph Warwick Bigger coined the term 'persister'. My research into Bigger hints, sadly, at him not being a particularly large man. Bigger was an Irish scientist who, in 1944, showed that penicillin kills only 99 per cent of a staphylococcus culture, leaving behind a stubborn 1 per cent that somehow survives. Bigger's cells weren't genetically different from the 99 per cent, but were what's termed 'phenotypic variants' – cells behaving differently from the others despite having the same DNA. These persisters are the guys that make penicillin pretty rubbish at treating chronic infections, such as the biofilms that form on implanted devices or infections of the cystic fibrosis lung.

The other day I was looking back at some of my undergraduate biochemistry notes (before burning them). They could have been written by someone else. In addition to an odd obsession with doodling surprised peacocks on every other page, undergraduate-Kathryn had written all these notes on subjects that, 10 years on, are completely alien to me. Amino acid catabolic pathways? Thermodynamics of enzyme interactions? Nope, all gone; replaced by the lyrics to every nursery rhyme ever written and a photographic memory of the film *Frozen*. Once, I probably saw *t*-test equations, but no more. Yet, as I flicked through all those peacock doodles, I found something that I still recognised. Biphasic kill curves. I don't know why this one little piece of knowledge stuck, but it did. Whatever the reason, I remember those biphasic curves to this day. A kill curve is a graph of death. You've got the number of survivors on the Y-axis and time on the X-axis. You add the weapon of death at the start, and the curve drops from its starting point at 100 per cent and keeps going down as all those poor bacteria succumb. And then something happens. The line starts to level off like the tail end of a ski slope. The rate of death slows, and while the bacteria are still dying, they're taking their time about it. Same thing when you treat real-life infections. In mice infected with TB, isoniazid causes a relatively rapid decrease in the number of bacteria in the lungs over the first four weeks. And then the killing grinds to a halt. You're left with maybe 1/1000 or 1/100 of the initial population, and they take much, much longer to die.

The reason is that not every cell is doing the same thing. A laboratory culture or a mouse lung is not a homogenous mixture of identical individuals. Like a human city, I suppose … a human city made up almost entirely of identical clones. Everyone is the same on paper, but when it comes to how they live their lives, not everyone follows the same path. Scientists call this phenotypic heterogeneity. I read one theory suggesting that when the going is good, a population of cells tends to behave in roughly the same way barring a very small number of rebels. But when life gets harder, the differences start to emerge. I can see similarities in the way the bacteria prepare for potential catastrophes and how today's global situation – economic depressions, terrorism, climate change – has been setting more and more of us humans on edge. The rise of Jeff the human survivalist has accompanied a general inkling of something bad being on the horizon (most likely a product of the media and internet giving us all a front-row seat in all the world's horrors). Like us, when microbes start to see the warning signs of impending doom, more and more of them change their behaviour as a survival strategy. A study looking at *M. tuberculosis* isolated from human sputum demonstrates this nicely. Galina Mukamolova's group in Leicester showed that bacteria adapted to life within the lung are more tolerant to first-line TB drugs than cells grown in the lab. A programmed adaptation to the intracellular environment most likely ensures that at least some of the bacteria are prepared for survival in the face of future trials and tribulations. A side effect of this adaptation is that the cells are also protected from being killed by antibiotics.

In 2015, a paper from John McKinney's group in Switzerland looked at what happens to populations of *M. tuberculosis* when they hit upon hard times. McKinney is part of a newish wave of researchers who believe that treating microbial populations as a collection of identical individuals is like trying to study global politics from space. By engineering *M. tuberculosis* to make green fluorescent protein during periods of active growth – the more growth, the more fluorescence – McKinney's group were able to visualise different populations of bacteria based on growth rate. In a culture flask, they found that everything looked fairly similar in terms of greenness. But when they exposed the bacteria to nutrient starvation, the diversity started to increase. Starvation, though, is the equivalent of a countrywide Twinkie shortage. A disaster, but nothing on the scale of the zombie apocalypse. So McKinney's team decided to test their bacteria under the most extreme conditions they were ever going to experience – a real-life infection where they'd have not only nutrient depletion to contend with but also attack from the host's immune system. When they looked at the bacteria clinging on to life in mouse lungs, they saw a huge range of phenotypic heterogeneity. Some cells were growing actively, others were ticking over slowly. A subpopulation of cells was non-growing but still metabolically active. Alive, but only just. These cells were absent in mice engineered to be lacking in a key component of the immune system, indicating that the formation of these non-growing cells requires immune challenge.

More recently, Sam Sampson's group in South Africa employed a similar technique, only this time they engineered their *M. tuberculosis* to make a fluorescent protein that is gradually lost from the cells as they divide. Sam infected macrophages with this strain and looked at how the population of cells diversified over time. While at early time-points the cells were all fairly similar, after 48 hours the scientists started to see a slow-growing population appear. As the macrophage infection progressed, an increasing number of bacteria also became tolerant to antibiotic challenge. The big question here is whether the slow-growing population is the same bunch of bacteria that can survive drug treatment. It would certainly make sense, but non-growth isn't the only path to antibiotic tolerance. Even if, for years, this was thought to be the case. Most antibiotics, after all, work on actively dividing cells. It's like, if an antibiotic is the spanner in the works for a bacterial cell, then switching off all your machinery means there's nothing to get snarled up. By choosing to go to sleep, the bacteria take away the weakness that the antibiotic could exploit. The spanner clatters through the stationary cogs and falls out the bottom. This idea of non-growing cells equals antibiotic tolerance is an important one. And as we saw in an earlier chapter, latent TB has traditionally been associated with the idea of hibernating bacteria. Remember Huber the Tuber and his copy of *The Magic Mountain*? It follows that one of the reasons why TB is so hard to treat is partly down to non-growing persisters protected from death.

A 2011 paper from Kim Lewis's group in the States looked at the genes switched on in *M. tuberculosis* persisters. This population was selected for by exposure to the antibiotic d-cycloserine. Once all the susceptible bacteria have died you're left with just the antibiotic persisters. Among the genes upregulated in Lewis's persisters were 10 toxin-antitoxin modules. Toxin-antitoxin systems are a yin and yang of the bacterial world, or a supervillain and his or her heroic nemesis. One protein – the toxin – inhibits vital functions within the cell; the other – the antitoxin – neutralises the toxin. By shifting the balance of the two proteins, a cell can regulate its own growth and shift into a non-growing state if required. *M. tuberculosis*, interestingly, contains a whopping 79 toxin-antitoxin pairs, suggesting that it probably uses them for something very important. In a more recent US–Korea collaboration, Lewis took a different approach and isolated mutants of *M. tuberculosis* with an increased ability to form persisters. The team started with a library containing mutants in every gene that could be mutated, then treated the mixture with antibiotics. If the mutant yielded fewer persisters, the mutated gene was vital for persister formation. If it generated more persisters – a super-persister – then the mutated gene was inhibiting persister formation. The genes identified came in a variety of flavours, including machinery involved in PDIM biosynthesis, phospholipid biosynthesis, metabolic pathways, RNA maturation and various other things. When they looked at the genes switched on in the super-persister mutants once again they found a number of toxin–antitoxin modules. In addition, these were genes involved in lipid biosynthesis, carbon metabolism and sundry transcriptional regulators. A bit of everything, basically. All of this suggests that *M. tuberculosis* persister formation involves multiple pathways containing lots of redundancy – different pathways and proteins that overlap in function (the 79 toxin–antitoxin systems are a case in point).

In parallel to his work on understanding the mechanisms of tolerance, Kim Lewis is also interested in methods of killing persister cells. But how do you kill something with a dozen or more backup options? 'We've hit upon a very interesting apparently Achilles heel not only in TB but in bacteria in general, and that is the Clp protease,' he told me. His drug, lassomycin, forces this protease to break down adenosine triphosphate (ATP), the energy currency of life. No matter whether a cell is growing, hibernating or something in between, it still needs ATP. Even switching everything off can't protect a cell from a drug that interferes with ATP. Kim told me how lassomycin – an antibiotic isolated from a soil bacterium – probably evolved specifically to kill persisters. A way for one species of bacteria to dig itself a niche in the overcrowded soil by killing its neighbours. 'This tells us that nature isn't targeting any mechanism of persister formation, probably because it's impossible due to redundancy. [Lassomycin] bypasses the complexity and it simply does something entirely different to hit persisters.' In a 2014 paper from Lewis's lab, there's one of those biphasic kill curves I mentioned above. Instead of the ski-slope curve, lassomycin results in a straight downward line, suggesting that there is no obvious persister population with tolerance to this drug.

Kim Lewis is a strong supporter of the idea that non-growth is behind the persister phenomenon. Some members of the TB research community don't agree. For the 60 or so years following Bigger's work, his idea that persisters were non-replicating or dormant cells persisted. It made sense, so few challenged it. It took until 2004 for someone to finally provide some evidence to directly support the hypothesis. But it wasn't all-species, all-antibiotics evidence. According to some scientists, switching off and shutting down is not a universal mechanism of persister formation. John McKinney has argued against what he describes as a scientific dogma arising from what was only ever a hypothesis. In one opinion piece, McKinney quotes Sherlock Holmes: 'It is a capital mistake to theorise before one has data. Insensibly, one begins to twist facts to suit theories, instead of theories to suit facts.' The big issue with straightening out these differences in opinion is that persisters are ridiculously hard to work on. It's like trying to study human Jeffs using population records for an entire city. These reclusive bet-hedgers are not only extremely rare among all the other non-Jeffs, but they don't behave in a way that makes them easy to observe. They're off the grid, lost in a sea of noise and avoiding doing the things that everyone else does, waiting patiently instead for the rest of their species to die so they can be proven right.

So, it actually wasn't until fairly recently that science caught up with persisters and came up with ways of spying on them. Microfluidics is sometimes labelled as lab-on-a-chip technology. It involves culturing bacteria in extremely small volumes in a miniaturised device that can be visualised using time-lapse microscopy. Because you're using tiny, tiny numbers of cells, you can watch what each one gets up to as an individual, rather than basing your conclusions on averages and generalisations. You can start out with, say, 100 cells, then add an antibiotic and watch 99 of those cells die. The one survivor is your persister, trapped in the chip's single focal plane with all its tricks laid bare. Another persister researcher, Sarah Fortune, says of single-cell imaging that 'there's an appealing concreteness to looking at cells and *seeing* that they're different from each other'. I have to agree with her on this. Kill curves or microarrays or a myriad of other techniques don't give you this same window into the secret lives of individual bacteria.

I've spent the morning watching some of John McKinney's time-lapse videos of persisters. In the middle of a dark field of vision, a single cell glows green. It's a Matrix-style set-up, with everything the cell needs to survive being pumped into the chamber and all its waste products being washed out the other side. The ideal conditions for our glowing cell to grow and divide, which is exactly what it does. Slowly, the cell lengthens and pinches off into two daughter cells that stretch away like the creeping roots of a tree, still touching end-to-end until the next division pushes them apart. Then an antibiotic – isoniazid – is added, and the cells stop growing. A little while later, the lights start to go out. One, then another. The green glows vanish to leave behind a dark shadow in the background. This shadow is the empty husk of a dead cell. The other cells keep on dying until all that are left are two determined green blobs. They don't seem particularly bothered about all the isoniazid; in fact, one of them continues to divide. From a distance, you see only in averages. It takes a technique like microfluidics to reveal that not every cell follows the herd. Most die. Some can survive. A few thrive.

The game-changing part of these experiments, though, was the observation that the drug-tolerant cells were not pre-existing non-growing cells. They weren't Jeffs who had shut down and switched off in preparation for an unseen apocalypse. They were not dormant, as 'dogma' would have us believe, but normal, growing members of the population. Something else was at play. I clicked onto the next video. This one flashed red like a Christmas tree. Flashing is not something I would ever have expected from bacteria. The red glow was coming from a labelled protein called KatG, or catalase-peroxidase. It's a mycobacterial protein hijacked by isoniazid to convert the drug into its active form. McKinney therefore hypothesised that the levels of KatG within a cell could be correlated with its likelihood of being killed by isoniazid. The flashing – or pulsing – was a surprise, though. It turned out that cells make KatG in short bursts, and it's during these bursts that it is sensitive to isoniazid killing. Once again, these results suggest that there is no single mechanism for bacteria to become tolerant to antibiotics. Some cells might be of the non-growing, dormant variety; others the result of stochastic gene expression as seen in McKinney's videos. Kim Lewis, however, would disagree with McKinney's assertion that his isoniazid-tolerant cells are persisters in the truest sense of the word. This is more an argument over terminology than science. Lewis uses the term 'stochastic resistance'; McKinney hijacks 'persister'. But, either way, all of it adds up to a lot of reasons why TB can be extremely difficult to treat in the real world.

A 2012 paper from Sarah Fortune's lab at Harvard also demonstrates that non-growing persisters are not the only solution to surviving antibiotic treatment. After all, when the zombie apocalypse hits, you don't need to go full-Jeff to survive. You might just be a faster runner than your neighbours, for example. Same goes for *M. tuberculosis*. When mycobacteria divide, not all the daughter cells are created equal. Instead, asymmetrical division and ageing of mycobacterial cells can lead to distinct subpopulations of cells with variations in growth rate and antibiotic susceptibility. Sarah's work showed that the slower elongating progeny are more sensitive to rifampicin than their faster counterparts. Isoniazid, on the other hand, preferentially kills the faster cells. Sarah's conclusions are that this subtle diversification may contribute to the variable outcomes of TB infection and treatment. She refers to these multiple layers of complexity as 'an incredibly tapestry of variation in the bacterial population'. Sarah is a fan of McKinney's work and the idea that there are all these cells behaving differently within a population. It's similar to JoAnne Flynn's findings regarding the granuloma being a dynamic, mini-state of its own within Lungland. Every granuloma is unique because it's made up of countless individuals.

At one end of the spectrum, there are the cells that slow down or stop growing. The small numbers of Jeffs with their bet-hedging strategies, or the larger numbers of cells forced into a non-growing state by hypoxia or nutrient limitation or low pH or exposure to toxic products of the immune system. Along similar lines to both of these, we have the formation of biofilms – structures of clumped-together bacteria. Think plaque on your teeth, only in your lungs. Not only are the cells within a biofilm not growing or growing slowly, but they're protected from antibiotics simply by being hidden away under lots of other cells. At the other end of the 'every cell is an individual' spectrum, there's all those stochastic differences subtly influencing antibiotic susceptibility. Asymmetric growth, for example, or random differences in enzyme levels such as those seen in McKinney's KatG studies. There's also efflux pumps. Efflux pumps are the bouncers of the mycobacterial world, tipping toxic compounds outside before they can get into fights with the staff. When *M. tuberculosis* is exposed to antibiotics at a level not quite capable of killing the cell, it up-regulates these efflux pumps to decrease the intracellular concentration of the antibiotic. This mechanism doesn't provide the cell with as much protection as a drug persister cell achieves, or true genetic resistance in which a drug's target is changed permanently. But it does a good enough job to keep at least some of the population alive, providing the concentration of the antibiotic doesn't get too high.

All of the above get together with that thick cell wall and say: 'There'll be no simple seven-day course of penicillin for you, you're going to have to try much harder if you want to eradicate me.' One of the big challenges in the TB field is that treating the infection takes a stupidly long time – six to nine months for drug-susceptible infections and around two years for drug-resistant. This is in comparison with other difficult-to-treat bacterial infections that require several weeks of antibiotics. So, how do we develop drugs to kill the (almost) unkillable? We'll start with the old-fashioned method that gave us our first TB drug, streptomycin, back in the 1940s. I'll say here that I'm not going into a huge amount of detail about the history of TB drugs, as this has been covered by other people. I'm more interested in talking about the different techniques available for drug discovery. To cut a very long and complicated story short, streptomycin came from the soil. It wasn't a serendipitous discovery in the vein of Alexander Fleming's discovery of penicillin. Instead, a scientist called Selman Waksman spent much of his career painstakingly isolating and culturing around 10,000 species of microbes from various natural sources. A thousand of these species had antimicrobial properties. A hundred could make these antimicrobials in the quantities needed to attempt purification. Ten of these antimicrobials could be isolated. One proved sufficiently effective against bacteria. Streptomycin. It changed the treatment of TB forever.

Streptomycin happened to be the first drug ever subjected to a randomised clinical trial – the gold standard in the development of modern-day drugs. The trial was actually based on good old-fashioned British penny-pinching. While streptomycin had already shown promising results in some patients, it was very expensive and many of those benefitting from it had bought their own supply. The UK government didn't want their limited purchases of the drug to be tossed about without any thought; not when they'd spent so much money on it. And so a 15-month trial in which patients were allocated to bed rest or streptomycin plus bed rest was carried out. Among the 55 streptomycin patients, four died in the first six months of the study compared to 15 in the bed-rest-only group. A good start! But then the streptomycin-treated patients who had started to recover deteriorated just like the others. Their TB bacilli had evolved resistance to streptomycin and the drug was no longer working. These results led to the first combination therapy for TB using streptomycin and another new drug discovered soon after (or just before, depending on reports): para-aminosalicylic acid (PAS).

The discovery of PAS was interesting. Streptomycin's discovery can be considered as a kind of bottom-up method. Find a compound first and see if it kills the bacteria; work out how it kills later if so inclined. In the case of streptomycin, it took a few years after its introduction for scientists to work out that it was inhibiting protein synthesis, and even longer to pick apart exactly how it went about killing a cell. The discovery of PAS as a TB drug, though, followed a top-down approach. The scientist responsible – Jörgen Lehmann – started with the pathway that he wanted to inhibit and worked backward. This was an example of target-based drug discovery; something that a large chunk of my own career would be dedicated to, 50 years later. Lehmann's idea for PAS came from a paper showing that a compound called salicylic acid is very good at activating the metabolism of *M. tuberculosis*. The TB bacillus was using the salicylic acid, in simple terms, as a food source. Lehmann reasoned that if he were to alter the structure of salicylic acid it might inhibit the bacterium's metabolism instead of switching it on. And so, after a heroic effort by a medicinal chemist called Karl-Gustav Rosdahl in synthesising the compound, PAS was born. Thanks to both PAS and streptomycin, TB was no longer an incurable disease.

Today, neither PAS nor streptomycin is used as a front-line TB drug but both are still useful in the treatment of drug-resistant infections. They have been superseded by the introduction of four newer drugs (and when I say newer, they are still over 50 years old). Isoniazid was discovered in 1952, pyrazinamide in 1954, ethambutol in 1961 and rifampicin in 1963. These four drugs are today prescribed in combination, with isoniazid and rifampicin providing the cornerstones of the therapy. Upon its introduction, isoniazid was hailed a wonder drug for the treatment of TB – to the point that apparently patients were enticed into dancing down the halls of a New York hospital in response to the announcement. It was cheap, simple and, combined with streptomycin and PAS, extremely good at treating TB without drug resistance emerging. Pyrazinamide, however, is the one of the four that I find the most remarkable. It's quite an odd drug in that it doesn't kill growing cells but is instead active against non-growers. Take it away from the four-drug treatment regimen and the time taken to cure an infection is increased, underscoring the importance of being able to target non-growing populations when treating TB. Pyrazinamide targets the cell's protein-making machinery, meaning the sleeping cell is unable to wake back up. I won't go into detail about the discovery of pyrazinamide, isoniazid, rifampicin and ethambutol, other than to say they all came from the cell-based, bottom-up approach. Their origins differed but they all started with the same premise – here are some compounds, let's see if any of them can kill bacteria. It's the method that has yielded the vast majority of our current antibiotics. Then came the advent of genomics.

Genomics was meant to revolutionise drug discovery. The *M. tuberculosis* genome was published in 1998, and suddenly we had the entire blueprint for this killer pathogen. All we had to do was read it and understand it, and the bacterium's secrets would be ours. It did spawn an entire new era of TB research – genetic manipulation, transcriptomics, proteomics, structural genomics, comparative genomics. Nearly everything I have talked about in this book wouldn't have been possible without the DNA sequence of *M. tuberculosis*. But when it came to drug discovery, it wasn't the new future everyone had hoped for. The idea was that a scientist could start out with a gene that looked like it might make an important protein. Genetic manipulation techniques could confirm that the bacterium really can't live without the target protein. From there, it was just a case of developing a drug capable of inhibiting the protein. This process involves first making and purifying the protein and finding a way to measure its activity. You can then use a gigantic robot to seed 364-well plates, toss in a huge number of potential inhibitors and see if any of them stops the protein from working. These inhibitors might be natural products or compounds that have been synthesised based on existing compounds, often as a result of other drug-discovery endeavours (for example, the hunt for cancer therapeutics). Once you find one that works, you can fiddle with its structure to hopefully improve its activity and other properties required for it to be a good drug (such as its toxicity against humans, how well it is absorbed into the body and its stability).

Of course, at some point you need to check that it actually kills bacteria. This is where I came in during my first postdoc position straight out of my PhD. I was the sole microbiologist among a massive team of chemists and biologists, all of whom had put an extraordinary amount of work into developing assays and using these assays to find compounds capable of inhibiting our protein of interest. I wonder to this day if they all fully grasped that my complete inability to keep up with them wasn't entirely my fault. *M. tuberculosis*, you see, works on an entirely different timescale to purified proteins. The pipeline went something like this: the chemists would give me a load of potential inhibitors in several 96-well plates. I would get back to them and say, no, screening this many compounds isn't possible, please can you narrow them down? With sad chemist faces, they'd reluctantly give me a list of the ones they were most interested in, and I'd tootle off to the TB lab. The chemists had a robot; when it came to the microbiology, I was that robot. I'd dilute all those compounds into some 96-well plates, add in a small amount of *M. tuberculosis*, leave the plates for around a week, then look at which wells had failed to grow. I'm not including here all the 'how to prevent Kathryn from dying' considerations that went into working on TB or the occasional issue of contamination (when your lab is set up to protect the scientist and not the sample, it's possible for random moulds or microbes to sneak into a culture).

Just to keep things exciting, my cultures would occasionally simply decide to not grow on their normal timescale and I'd have to start over. Another time, a fly got into the lab and we had to have no end of meetings discussing whether The Fly was going to get some *M. tuberculosis* on its feet and carry it out the door to infect some poor, innocent child or maybe the entire world. And one time, the lid came off someone's culture while it was in the incubator and we all could have died but didn't because of all the meetings that had prepared us for this very eventuality and it was actually fine, but we had a lot more meetings to discuss how we could stop it from happening again and the safety department went and threw some of our water-filled culture flasks off the roof to make sure they wouldn't break if dropped. Anyway, the whole process of testing some potential inhibitors against *M. tuberculosis* – if it went to plan – took maybe nine days or so, during which time the chemists would no doubt have made an additional 9,856 compounds, of which I might be able to test maybe 100.

Let's summarise the entire, extremely expensive and time-consuming project: it didn't work. Our compounds were brilliant at inhibiting the protein by itself. When it came to *M. tuberculosis*, though, it only died a little bit, and definitely not enough for the compounds to be of any use as new drugs. I tried to work out why. Was it the super-thick cell wall stopping the compounds from getting inside? Was the bacterium pumping the compounds back out? Did the compounds work better when the bacterium was contained inside a macrophage? Was the target not actually that good a target? And then we gave up and moved on with our lives.

We weren't the only ones. At some point between me joining this project and its unhappy demise, the prevailing view in the TB drug-discovery world shifted. Target-based approaches like ours went out of fashion and many people started saying how cell-based approaches like the 'good old days' were the way forward. *M. tuberculosis* was just too complicated to rely on a compound targeting just one enzyme. We needed the multi-target activities of drugs such as isoniazid, and compounds that we knew were capable of getting inside a cell right from the first screen. It's not just TB drugs that have suffered from target-based drug discovery failing to live up to its promise. There's a really good paper from 2007 summarising GlaxoSmithKline's experiences looking at more than 300 genes and 70 high-throughput screens over the course of seven years searching for antibacterial agents. GSK is a massive pharmaceutical company. They have robots, not Kathryns, and these robots screened more compounds that I could have investigated in an entire lifetime. And still, all their efforts yielded a grand total of no new antibiotics.

According to Kim Lewis, much of the issue was the result of the collapse of the successful antibiotic drug-discovery platform that yielded many of our current drugs between the 1940s and 1960s through the mining of soil organisms. The problem was that only 1 per cent of soil microbes can be cultured on a Petri dish; the majority don't grow. Kim told me that once that 1 per cent was exhausted, no new discovery platform emerged to replace it. So while the few drugs to have been discovered since the 1960s have used a cell-based screening approach, this doesn't mean that this method isn't just as failure-prone as my target-based attempts. The 'good old days' are gone, and it's not as simple as going back to basics. 'All the new compounds that you see in all fields including TB are a complete lottery,' Kim explained. 'One person has a lucky strike and you have one compound. And then that either succeeds or fails, and it usually fails due to numbers – the probability of failure is higher. Without a platform, we are doomed.' His view is that we can't keep up with the need for new antimicrobial drugs using this hit-and-miss strategy where the misses far outnumber the hits and any success is mostly down to luck. I should say that my view of the current TB drug-discovery pipeline is that it has come a very long way in recent years. I'll talk about some of the success stories in the final chapter. For now, though, I don't want to ruin the general feeling of pessimism and misery I have cultivated in this chapter. You're welcome.

In summary? *M. tuberculosis* is very difficult to kill, most of our current drugs are over 50 years old and, even when drug resistance isn't present, these drugs still don't do a good enough job of targeting every cell within a population. What we really need are new treatment regimens that work faster than the existing combinations of drugs. For this to happen, we need new drugs that can target persisters of all flavours. No more biphasic kill curves, no more efflux-mediated tolerance and hopefully no more latent tuberculosis. One of the key issues is that while we know that persisters exist in a range of different types, no one knows exactly what the therapeutic relevance of each type is. So straight away, screening for new drugs capable of killing persisters is difficult, especially as most drug-discovery screens are performed against growing bacteria. Even if you find a drug that can kill, for example, hypoxic TB in the lab, will this translate into sterilising a real-life infection? And would killing these cells do anything to cure TB? An intriguing approach could be to target the cell membrane rather than specific pathways involved in persister formation. It doesn't really matter why and how a cell has become a persister – poke a hole in its membrane and it's going to die.

We talk about drug resistance as one of the biggest threats to TB control, but persistence is a threat that's always been there. Persistence and resistance work together in a vicious cycle. Without drugs that can kill TB faster than at present, drug-tolerant populations of cells can help provide an infection with the opportunity for genetic resistance to emerge. In turn, resistance puts more pressure on a drug–development pipeline that still hasn't recovered from its 50-year slump. What we need is something new and transformative, otherwise we will be giving *M. tuberculosis* the opportunity to evolve into a drug-resistant monster that can no longer be treated by 'modern' medicine.

CHAPTER TEN

The Drugs Don't Work

By the time doctors figured out that Phumeza Tisile's TB infection was resistant to kanamycin, she was already deaf. Diagnosed with TB at 19, South African Phumeza was prescribed the standard four front-line drugs. After five months her condition was still deteriorating, so she was started on treatment for MDR-TB, including a drug called kanamycin. It's a painful injection, taken daily, with a variety of potential side effects. For Phumeza, the drug robbed her of her hearing while doing nothing to help cure her TB. Her infection was what's known as extensively drug-resistant – or XDR-TB. If she'd been diagnosed correctly to start with, she could have avoided months of ineffective treatment that cost her not only her hearing but nearly her life.

'The treatment for MDR and XDR is horrible; it makes you even sicker than you already are,' Phumeza told me. 'I remember almost every day I would vomit after taking the tablets.' Can you imagine that? Taking around 20 tablets a day and injections for six months, knowing that each dose is going to make you throw up; swallowing your own vomit so that you don't have to repeat the dose? In her blog for Doctors Without Borders, chronicling her experiences with drug-resistant TB, Phumeza says: 'You sit down by yourself and think: "If this is a dare, bring it on. I'll do whatever it takes just to survive." I did not want to be another statistic of TB. Yeah, sure, one day I will die – but not from this horrible disease.' Until I read Phumeza's blog, I'd never thought about TB as a fight waged not only in a patient's lungs, but in their heads and hearts as well.

Phumeza had just started a course at university when she became ill. She was losing weight but not sweating or coughing, so it was a shock when TB was discovered on a chest X-ray. It was the start of a 20,000-pill odyssey that would see her spend a year in hospital, lose fellow patients and friends along the way to both TB and suicide and leave her with a permanent disability in the form of hearing loss. At one point early in her treatment, she became so sick that she was moved to the Karl Bremer Hospital. She remembers being left alone in a dark room, musing that perhaps the doctors thought she was already dead. It's this loneliness that strikes me the hardest. The idea that surviving drug-resistant TB is a very personal fight that can see a patient isolated from their friends and family, with a mask keeping them at arms' length during any face-to-face interactions, and the sheer enormity of what is involved in curing the disease almost inconceivable to someone who hasn't gone through it.

Drug-susceptible TB is treated with a two-month intensive phase of a four-drug cocktail: isoniazid, rifampicin, pyrazin-amide and ethambutol. This is followed by four months of just isoniazid and rifampicin. Side effects include nausea, vomiting, liver damage and reddish-orange tears and sweat. That's for the 'easy' cases. But for a growing number of patients, this standard regimen doesn't work due to drug resistance. MDR-TB is caused by strains of *M. tuberculosis* that are resistant to at least isoniazid and rifampicin – the two cornerstones of TB therapy. But that's OK. There are other drugs that can be used; they're just not as good and come with a variety of unpleasant side effects such as hallucinations, psychosis, blurred visions, skin problems, liver and kidney damage and hearing loss. Then we have XDR-TB. It is the Kim Jong-un to MDR-TB's Kim Jong-il. This slightly more lethal descendant has the obligatory resistance to isoniazid and rifampicin plus a fluoroquinolone and an injectable agent. (There's been some question about whether XDR-TB deserves its own name when it's really part of the MDR-TB spectrum. I like it, though, as it makes people sit up and take notice of the problem.)

Depending on what drugs the strain of *M. tuberculosis* is resistant to, a four- to seven-drug treatment regimen is chosen from whatever is left. First, you've got your injectable agents such as streptomycin (the first TB drug ever introduced) and the fluoroquinolone family. Once you exhaust those options, you're left with the less potent and often poorly tolerated old-school drugs as well as newer agents that have yet to fully prove themselves. Phumeza was lucky that her doctors were able to treat her with off-label linezolid, a drug not initially developed for the treatment of TB. Treatment duration for XDR-TB generally involves an eight-month intensive phase involving daily injections followed by a full year, more or less, in which some of the drugs may be stopped depending on how a patient is coping. Phumeza describes the drugs as 'bastards going down my throat'. She had to keep taking them for two years, despite the side effects of linezolid alone meaning that its official, labelled use (for example, in treating MRSA skin infections) is limited to 28 days.

Doctors Without Borders released some brilliant infographics this World TB Day (2016), showing the huge burden of pills that an MDR-TB patient has to swallow. According to them, it can take 14,600 pills to treat just one person. Stacked up end-to-end, they could make a tower as high as the Golden Gate Bridge. For an uncomplicated drug-resistant case, a course of these drugs costs between $1,800 and $4,600, but this doesn't cover the newer drugs and it doesn't include all the costs of caring for a patient such as long-term hospital stays and dealing with side effects. Once you get to XDR-TB, the price tag skyrockets. Depending on an individual's resistance profile, the severity of their disease, co-morbidities and where they live in the world, the cost of their treatment can push $1 million (Australia). Of course, if you're 'lucky' enough to be diagnosed with XDR-TB in a country such as Australia, or the US or UK, then you'll have the benefit of a whole team of doctors doing their best to cure you using the best diagnostics and drugs in the world. I'm not promising that the human side of things will be all that fun; they may well focus mainly on the medical problem while your physiological state shrivels up in a tiny isolation room. But you'll most likely live, so that's a good thing. Anyway, the point I'm trying to make is that regions like KwaZulu-Natal, South Africa, have far more cases of XDR-TB than Australia or the US or Western Europe, and far less money to spend saving individual patients. This is the crux of the matter. For someone with TB, it doesn't matter that they're one of 10.4 million active cases per year. Each patient isn't a data point on an incidence chart, they're an individual fighting an individual battle against the disease. It's all well and good saying that current diagnostics catch the majority of cases, or standardised treatment has saved however many lives. But if you're one of the people who is missed, then who cares that overall mortality rates are going down?

Phumeza was lucky that Doctors Without Borders were able to come to her rescue after she was diagnosed with XDR-TB. Doctors Without Borders is an independent association that works in a range of settings to provide impartial medical aid. They've been fighting TB for over 30 years and in 2014 treated 23,300 patients with first- or second-line drugs. Through a programme to provide individually tailored treatment for XDR-TB, they arranged for Phumeza to receive both linezolid and a second expensive drug from abroad, clofazimine. While the bulk of Phumeza's treatment was funded by the South African government, linezolid in particular is just too expensive to be part of their standard regimen for XDR-TB. At the time, the drug was priced at over $40,000 per patient when purchased through the private sector. But without this drug Phumeza would almost certainly have died. Finally, three years after she was first diagnosed, she became one of the first South Africans to be cured of XDR-TB. The staff, doctors and fellow patients at the Lizo Nobanda TB Care Centre celebrated too, presenting Phumeza with a framed certificate featuring a photo of her standing on a mountain with her arms spread wide. She had conquered her infection despite the awful odds stacked against her. In 2012, only a quarter of XDR patients were successfully treated.

XDR-TB is a disease that only really appeared on the world media's radar a little over 10 years ago. I remember the news reports about this sudden manifestation of a drug-resistance nightmare that had seemingly appeared out of nowhere. Why hadn't we done more sooner, everyone asked? It seemed almost unbelievable to me that we'd finally reached a place where TB was almost incurable. 'Everybody knows that pestilences have a way of recurring in the world; yet somehow we find it hard to

believe in ones that crash down on our heads from a blue sky,'
says Albert Camus in *The Plague*. It was how it felt to hear about
XDR-TB as a relative outsider, having only just begun working
in the TB field myself. I now realise that there were never any
blue skies overhead, and in the case of XDR-TB we really
shouldn't have been surprised when it came crashing down. It
was in fact a problem that had been brewing for years. It just took
someone looking in the right place to notice what was happening.

Gerald Friedland has stood on the cutting edge of TB and
HIV medicine for a large chunk of his career, including
working to bring the 1980s Bronx TB epidemic back under
control. His theory is all about how we should be learning
from epidemics like this, which when you take a look at the
contributing factors, were inevitable. No falling from blue
skies here. The New York TB outbreak and, later, XDR-TB
were dark clouds looming on the horizon for many years
before we started to feel the rain. In 2003, Gerald took his
pioneering approach to HIV/TB treatment integration to
South Africa. There, HIV notification rates were 950 per
100,000 people compared to the US's 3/100,000. TB was
similarly high, reaching 1,000/100,000 people in the region
chosen for the study – Msinga in the district of KwaZulu-
Natal, around 2.5 hours north of Durban. Here, like many
other places in sub-Saharan Africa and all across the world,
TB and HIV strike together. Msinga is home to 180,000
Zulu people and is the poorest sub-district in South Africa.
The Church of Scotland Hospital in the town of Tugela Ferry
was chosen for Gerald's study by virtue of the nearby location
of a TB drug-resistance testing lab and the presence of a local
NGO. With the help of YouTube, I take a drive into Tugela
Ferry along a rock-strewn road shaded by crumbling
mountains on one side and a red and dusty ravine on the
other. A small group of goats nibble at parched bushes on the
verge as traditional Zulu dwellings and the mismatch of
small, square buildings pop up among the rocks. The hospital
we're heading for had its origins with nineteenth-century
missionaries. The Gordon Memorial Mission, named after a
missionary who drowned at sea before even reaching South
Africa, set up schools and a clinic. Then in 1939 the hospital

was moved closer to the Tugela River, and the Church of Scotland Hospital was born. From 2003 to 2006, Gerald and his colleagues worked on improving the hospital's diagnosis of TB and HIV, and implemented new methods of ensuring patients got the right treatment for both diseases. One such intervention was based on the observation that the Zulu culture is very much based around working together collectively, so the team implemented an Adherence Club. Every week, patients sit together dividing out the many, many pills that need to be taken several times a day. It's a simple thing but it makes a big difference.

However, these interventions didn't have as great an impact as expected. With 12 per cent of patients still dying, the team looked closer at the strains of TB responsible. All of those who died had a highly drug-resistant strain that fitted the recent description of XDR-TB. In fact, of all the TB cases that had occurred during 2005 and the beginning of 2006, 221 patients were retrospectively found to have had MDR-TB, 53 of whom would have met the definition for XDR-TB. Only 55 per cent of these patients had been previously treated for TB, meaning that their drug resistance wasn't the result of poor compliance or incorrect treatment. They'd picked the strain up from someone else, probably as a result of the long in-patient stays at the hospital. They were coming in for one thing and leaving with XDR-TB. This was different from the previous, isolated cases of XDR-TB that had popped up in various parts of the world. This strain was spreading from patient to patient at an alarming rate. The part of this outbreak that caught everyone's attention was that, of the 53 XDR patients, 52 had died. The media went crazy, with one newspaper dubbing Tugela Ferry a 'Town of Death' and the world's news organisations spreading the word that almost untreatable TB was upon us. *Run for the hills, everyone, we're all doomed.* That sort of thing. I can remember family members who, knowing I worked on TB, asked me if we were at risk over in the UK. Was this the dawn of a new apocalypse, in which XDR-TB would creep aboard every plane and spread across the globe? Maybe someone should be locking up all those XDR-TB patients to stop them from coming in contact with others (we'll come back to this ethical minefield later)!

Obviously, we didn't all die, and hopefully the irresponsible journalists responsible for the whole Town of Death thing feel pretty damn guilty now. The outbreak proved disastrous for the local economy, as no one wanted to go there. Yet the reality was that Tugela Ferry was little different from many other towns in KwaZulu-Natal. It just happened that Gerald's team was looking closely enough to notice what was happening in the Church of Scotland Hospital. People were dying of XDR-TB everywhere else too, only their deaths were going undiagnosed and unreported. The disease has now been identified in all regions of the world, but these outbreak clusters in the KwaZulu-Natal province are still a big worry. Between 2011 and 2014, there were around 1,000 XDR-TB cases detected in this one region. The true number is likely to be much higher as a result of a lack of diagnosis. It's scary to think that XDR-TB patients may be dying of what doctors believe to be an easily treatable infection, and their deaths explained away as the result of poor compliance with treatment. Or they're dying without ever being diagnosed in the first place.

It's not all doom and gloom, though. XDR-TB, as stories like Phumeza's show, isn't a death sentence and outbreaks can be prevented to some extent. In 2009, a paper authored by Sanjay Basu and featuring Gerald Friedland among the contributors set out a series of control measures to prevent epidemics of XDR-TB from arising. Looking back, the problems in Tugela Ferry were predictable based on the perfect storm of risk factors surrounding the town and others like it. There was a high TB case rate combined with a weak diagnostics and control programme, all thanks to the extreme poverty in this region. Then HIV arrived on the scene in the 1980s and the TB rates rocketed. Patients were routinely hospitalised (rather than treated in the community), but the stuffy, crowded wards were better suited to spreading disease than curing it. Everything was set for a drug-resistant strain to emerge and, importantly, to go unnoticed as it jumped from person to person. So preventing or reversing an outbreak can be greatly aided by addressing these risk factors through active case-finding and early drug-resistance testing to catch an outbreak before it takes hold, community-based treatment rather than hospital admission, and simple healthcare interventions, such as opening windows and directing potential TB patients into separate waiting rooms. All these changes to how TB is dealt with in places like Tugela Ferry have helped, but XDR-TB is still out there. It's going to take big improvements in diagnostics, treatment options and healthcare availability to fully solve the problem. With only one in five people worldwide receiving the treatment they need, the problem of drug resistance is not going anywhere anytime soon.

In 2015, the ongoing tide of genome sequencing advancements meant that scientists could, for the first time, look into how XDR-TB outbreaks emerge. A group led by Ashlee Earl compared 337 clinical isolates collected from all 11 districts of KwaZulu-Natal over six years. All of the strains were highly related, so the outbreak was definitely due to transmission between patients rather than independent emergence of drug resistance in multiple individuals. What was really interesting, though, was that an evolutionary tree based on when the strains diverged from one another suggested that the XDR-TB outbreak was four decades in the making. As far back as the 1950s, initial strains resistant to isoniazid and streptomycin were beginning to emerge. Come the 1980s, MDR-TB also resistant to rifampicin and pyrazinamide was in circulation. Finally, in the 1990s, the strains accumulated kanamycin and ofloxacin resistance, spawning XDR-TB. The mutations were likely the result of gaps in TB-control efforts dating back some 50 years – poor antibiotic stewardship meaning that, as each new drug was introduced, resistance mutations rapidly became fixed in the population. While drug resistance in individual microbes is something that arises naturally, it's this misuse of antibiotics that allows resistant strains to multiply. When Alexander Fleming accepted his Nobel Prize in 1945 for the discovery of penicillin, he predicted this future we now live in:

It is not difficult to make microbes resistant to penicillin in the laboratory by exposing them to concentrations not sufficient to kill them, and the same thing has occasionally happened in the body. The time may come when penicillin can be bought by anyone in the shops. Then there is the danger that the ignorant man may easily under-dose himself and by exposing his microbes to non-lethal quantities of the drug make them resistant.

Among any population of bacteria there are mutants in which small changes in their DNA sequence make them subtly different from their companions. These gradually accumulating differences are the same differences that can be used to date the point at which two strains diverged. Sometimes, though, the mutations impair the microbe's ability to compete with the rest of the population. These mutants won't stick around for long, not unless something happens to give them an advantage. For example, a mutation that randomly alters the target of an antibiotic will most likely decrease the 'fitness' of the bacterium, meaning that it will struggle to survive. But in the presence of an antibiotic, this mutant will come into its own and flourish where every other microbe dies. A sole survivor who can go on to repopulate in its own image and overcome the original fitness cost with additional mutations. After the introduction of streptomycin to treat TB in the 1940s, resistant strains of *M. tuberculosis* rapidly emerged. When you consider the probabilities involved, it was an inevitability. Streptomycin-resistant mutants emerge in *M. tuberculosis* at a rate of approximately 1 in everyone 2×10^8 (two hundred million) cell divisions. In severe cavitary TB, a patient has something like 10^{12} (a trillion) bacteria dancing a merry jig in their lungs, dividing again and again. So potentially, a fair few of these bacteria will be naturally resistant to streptomycin. Mutants will emerge and disappear, and emerge again. But were you to challenge the infection with streptomycin alone, it would open the door for one of these resistant mutants to take over. The mutant divides and divides and divides until there are 10^{12} bacteria once again, all of them resistant to streptomycin. This is the reason why combination therapy for TB was introduced. Spontaneous resistance to one antibiotic is an easy task for *M. tuberculosis*. Two at the same time, though, is a very rare event. Multi-drug resistance emerges sequentially, often with a little help from incorrect usage of antibiotics. For example, something like 10 per cent of current TB infections in the US are already resistant to isoniazid. Standard treatment for TB includes four months in which a patient takes just rifampicin and isoniazid. If the initial isoniazid resistance is missed, any one of the spontaneous rifampicin mutants can rise up and – ta-da! – multidrug-resistant TB.

South Africa-based researcher Alex Pym was telling me how standardised treatment can sometimes inadvertently make the situation worse. 'Everyone gets the same group of drugs and that only gets modified two or three months into therapy,' he said. This matches up with what Phumeza told me about her own infection. When these front-line drugs fail, this is the point at which a sample is sent for drug-susceptibility testing. By the time the results are in, the patient may already have begun the standardised treatment for MDR-TB. 'You may start patients on treatment where they only get two drugs or so that are active against the infective organism,' Pym said. 'By the time you get to drug-susceptibility testing, you have more resistance.'

It's a similar situation in India, according to Dr Nerges Mistry, director of the Mumbai-based Foundation for Medical Research. 'For a long time standardised drug regimens were used for the treatment of drug-resistant patients. But in Mumbai particularly it was shown with evidence that at least 60 per cent of the patients were resistant to four of the six second-line drugs that were in use in the standardised drug regimen. That is why I think that the shift has taken place reasonably quickly within a period of the last 12 months or so to individualised drug testing.' This testing, however, isn't routinely performed until a patient fails the first-line drugs. 'In cases who have risk factors,' Mistry said, 'such as if they're contacts for MDR patients or they have HIV or diabetes, then it is recommended – but it doesn't always happen – that drug-susceptibility treatment is done right at beginning of even first-line treatment.' I'll come back to the reasons why drug-susceptibility testing is such a problem in TB in the chapter on diagnostics.

When I first started to become interested in TB aged 20-ish, I naively believed that drug resistance was an individual problem. After all, in the UK (and probably everywhere else) we are constantly told to make sure we finish a course of antibiotics. 'Take the full dose', the label on the packet reads. Otherwise we might not kill all of our throat infection or whatever, and it will be all our own fault when it comes back as a drug-resistant infection. So I concluded that drug-resistant TB worked in the same way and it was a patient's own responsibility to see their treatment through to the end. If people would only take all their pills, the problem would be solved! Nowadays, I can't help but think that there is too much emphasis placed on the role of the individual in preventing antibiotic resistance when, in reality, it's not as simple as this. A 2010 study looked at the emergence of MDR- and XDR-TB in a South African gold mine, which to all extents and purposes was doing everything right when it came to TB control. Patients were getting the right drugs based on WHO guidelines; they were monitored closely and treated on wards with exemplary infection control, and supervised as they took their medication. Adherence rates ranged from 95 per cent to 98 per cent – patients were sticking to their treatment regimens, there's no doubt about it. Only, during the study period, the strains of *M. tuberculosis* kept on accumulating new resistance mutations. It raises the question of whether adherence is a red herring when it comes to the evolution of resistance.

A 2011 paper was published in the *Journal of Infectious Diseases* suggesting that non-adherence alone is not sufficient for the emergence of drug resistance. The authors grew *M. tuberculosis* in the lab and then tried to kill it using the traditional combinations of drugs prescribed for drug-susceptible TB. When they stimulated non-adherence, they discovered that a huge 60 per cent of doses had to be missed before the therapy would fail to kill off all the bacteria. In no experiment did resistant populations of bacteria expand to more than 1 per cent of the total. Instead of non-compliance, it seems that, through a quirk of nature, a small proportion of patients just happen to deal with TB drugs differently from others. There's always going to be variability in how someone's body absorbs, distributes and breaks down a drug – a bit like how some people can cope with higher doses of alcohol whereas I'm giggling insanely after one glass. In some cases, this natural variability in the inner workings of a human body can lead to concentrations of one or more TB drug falling below the dose needed to kill *M. tuberculosis*. Sub-lethal dosing like this is worse for the emergence of resistance than stopping a course of antibiotics before the course is complete.

In 2012, there was a really interesting paper authored by Aurelia M. Schmalstieg, the owner of one of the best-sounding names in science. She describes a theory called the 'antibiotic resistance arrow of time', in which efflux pumps give bacteria the chance to evolve resistance mutations to more than one drug. The idea is that these pumps can keep a drug-sensitive bacterium going in the presence of not–quite–high–enough–to–kill concentrations of an antibiotic. It's the breathing room the cell needs to sequentially accumulate resistance mutations to two or more drugs. Using *Mycobacterium avium* (a species that can cause disease in HIV-positive individuals), the paper's authors tested what happens to mycobacteria exposed to different doses of antibiotics. Within three days of less-than-perfect treatment, there was a 56-fold increase in the expression of two efflux pumps, providing the cells with low-level resistance that protected some of them from being killed. This resistance isn't down to changes in the DNA – it's an entirely reversible phenotypic adaptation. I think of it as transient state in which the bacteria are on their way towards resistance but haven't quite made up their minds. If antibiotic treatment is stopped, they switch off the efflux pumps. If you continue to expose them to sub-lethal concentrations of antibiotic, resistance mutations emerge. In a real-life infection, these sub-lethal concentrations could result from natural quirks of a patient's drug metabolism, differences between how easily a drug penetrates an individual granuloma, poor-quality drugs or the wrong prescribed doses. The way round it is to fiddle with dosing schedules to ensure that antibiotic concentrations are maintained at a level against which efflux pumps cannot compete. But, in this scenario, no treatment would actually be better than insufficient treatment.

What all of this helps to reveal is that drug resistance is not a one-shot, all-or-nothing process. On a single-cell level, resistance might look like a lottery with very poor odds. Except a population of bacteria has billions of tickets and there's more than one prize to be won. The recent advances in genome sequencing have stimulated a growing appreciation for the dynamic nature of the process. Within a single patient, *M. tuberculosis* can test out various different drug-resistance mutations and pick the one with the lowest fitness cost. Along the way, there's plenty of room for false starts and experimentation. A 2015 paper from Qian Gao's group in China used CT imaging to visualise what was happening in the lungs of a patient with very extensive TB. Eight years previously the patient had supposedly been cured of TB. As a young adult, however, she became ill once again. The imaging of her lungs showed six isolated lesions, three of which were open to the airways and producing sputum. Drug-susceptibility testing on a culture isolated from her sputum revealed resistance to both isoniazid and rifampicin as well as to an injectable agent, amikacin. One step away from XDR-TB, in other words. Because of her history and the severity of her TB, she was started on an intensive treatment regimen using seven different drugs. However, a repeat CT examination after eight weeks of drug treatment showed that the different lesions were not all responding in the same way. While one had decreased in size, others had increased. It all suggested that the individual lesions might contain separate populations of bacteria with different drug sensitivities. So the paper's authors sequenced the bacteria from the patient's sputum and discovered not one but three sub-clones. It wasn't a mixed infection, as in three unrelated strains of TB infecting one patient. Instead, the strains were highly similar and likely evolved from the same parent. All three had taken subtly different routes towards drug resistance – an example of branched evolution.

As was the case for this unfortunate Chinese patient, previous treatment for TB is a huge risk factor for the development of a drug-resistant infection. One of the problems can be standardised treatment regimens used in the absence of drug-susceptibility testing. This has been a contributing factor in India, topping the tables for the highest levels of drug-resistant TB in the world. With 2.2 million cases during 2015, India is the country with the biggest burden of TB. Around 40 per cent of the population of 1.25 billion people are believed to be latently infected, representing an extraordinary reservoir of potential disease. 'In Mumbai we've seen a rapid increase not only in the rates of drug resistance but also the severity of drug resistance,' Dr Mistry told me. India's surveillance system, however, has historically failed to accurately detect the true numbers of drug-resistant cases. Talking of some of her own work looking at drug resistance, she says, 'In new cases, it was as high as about 24 per cent, and in previously treated patients it was 34 to 35 per cent.' This differs greatly from the official reports that, at the time, put new case resistance at 1 to 2 per cent and treated cases at 5 per cent.

India's TB problem is multi-factorial: poor diagnostic infrastructure, lack of trained healthcare staff, crowded living conditions and various social issues, as well as high rates of type II diabetes (which triples the risk of TB) and smoking (which doubles the risk). The country has also suffered hugely from mismanagement of TB treatment by some parts of the private health sector (some practices are very good, others not so much). Stroll down the street in Chennai, and chances are you'll pass by one of the 550 registered hospitals or countless medical practitioners – licensed or unlicensed. It's the health capital of India, attracting hundreds of thousands of foreign patients hoping to take advantage of the low prices and latest technologies.

This embracing of modern medicine is, unfortunately, taking its time to trickle down from the gleaming heights of

Chennai's billion-dollar health-tourism industry to the everyday treatment of TB. Although the country's TB-control programme provides free TB healthcare, most patients initially approach private practitioners with their symptoms and half end up being treated outside the official channels. A 2014–2015 survey of 228 private practitioners from the city revealed that their adherence to India's TB-control programme's guidelines was not great. Patients with a long-term cough were rarely sent for TB laboratory testing despite this being a huge red flag for TB infection. Three-quarters of those surveyed used incorrect testing for patients suspected of having TB, two-thirds failed to monitor treatment response and 8 out of 10 didn't notify authorities of confirmed TB cases. Another 2013–2014 survey based in Chennai revealed that a quarter of private practitioners were unaware that TB notification is mandatory, and even among those aware of the government order, only a small proportion were following through. A lack of time, worries about patient confidentiality and concern about patient stigma have been reported as potential barriers against notification in a number of studies. But without these data, TB-control measures cannot be evaluated and improved.

The fragmented healthcare system in India is part of the issue. Patients may see a number of practitioners – ranging from pharmacists to traditional healers to public and private doctors – before a diagnosis is even made and treatment started. A 2014 systematic review noted delays in diagnosis ranging from 4 to 268 days. And these studies are probably reporting the better end of the spectrum, while some patients who approach unregistered healthcare providers with their symptoms don't get any further. On top of the delays, a 2015 systematic review of TB care in all of India (private and public) found that the treatment of TB was variable and often very poor. Almost all of the studies included in the review reported that less than half of healthcare providers knew what the correct treatment regimen for TB was. Compounded with the delays in (or a lack of) diagnostic testing, high levels of treatment failure and drug resistance are unsurprising.

'The private sector is an extremely heterogeneous sector,' Dr Mistry told me. 'You have the doctors who are completely unqualified practitioners [and] have their clinics in the worst parts of the city – the most vulnerable parts of the city. And of course the treatment given is often arbitrary and variable, and sometimes even downright irrational.' But Mistry doesn't think the private sector is at fault for India's problem with drug resistance. After all, it was the public sector that failed to notice the problem right under its nose, and it took a private doctor to finally force India to admit it had a problem.

In 2012, Zarir Udwadia – a Mumbai-based doctor – reported the first cases of totally drug-resistant (TDR) TB in India. All four of his patients were resistant to all the front-line drugs and all the second-line drugs tested. Looking back at their prescriptions revealed that three of the patients had received erratic and unsupervised treatment regimens from multiple private practitioners. Until 2010, India's free TB treatment was limited to susceptible and not drug-resistant TB, meaning that patients like these with MDR-TB were left with little choice but to resort to private practitioners. However, a Mumbai study of private practitioners in Dharavi revealed that only 5 of 106 were capable of prescribing the correct course of drugs for a hypothetical MDR-TB patient. Get it wrong and resistance is amplified – as seen in the cases of these TDR-TB patients. The resulting media storm made India and the rest of the world sit up and take notice. 'That really set the cat among the pigeons,' Dr Mistry told me. 'There was complete panic about the situation, and it actually drew attention to the problem of impending drug resistance in India.' The Indian government was not happy with Udwadia at first; something that Mistry puts down to the National TB Control Programme's politics of denial and the refusal to accept the scale of the TB problem. Udwadia effectively forced the government to face up to the reality of the situation, and it did go on to start investing more money in combatting TB. The situation is far from under control, though. Mistry describes India's battle with TB as 'technologically limping along in the right direction' but believes that unless socioeconomic changes are brought in to combat, in particular, poor housing, sanitation and overcrowding, then TB control in the country is mere firefighting. 'It is a dismal picture for the future,' she says, 'for while the country is placing an enormous emphasis on economic growth and so on, unless you have a healthy population and unless the determinants of that health are worked on, we're not going to go anywhere.'

The changes under way in India have come too late for the 15 TDR-TB patients. They were all started on salvage regimens made up from whatever drugs their infections were susceptible to, as well as surgical removal of infected parts of the lungs in some patients. Within less than a year of Udwadia's first TDR-TB paper being published, five had died. It got me thinking about what can be done for a patient with a TB infection untreatable by current medicine. Do you send them home to die and risk them infecting others, or keep them locked up in a hospital-prison for the rest of their life? This is the part of the drug-resistance problem where I start to really struggle. Is it ever right to take away someone's civil liberties in the name of public health? In Dustin Hoffman's 1995 film *Outbreak*, an Ebola-like virus threatens to escape from the Californian hub of infection, leading to a military and government plan to bomb the entire town. Operation Clean Sweep is avoided thanks to some superhuman science-ing and a dramatic helicopter-versus-bomber standoff. A rather extreme situation, I agree, but is it really a million miles away? I've been reading up on the South African XDR-TB epidemic that started with the events in Tugela Ferry, and the 2008 news reports read like something out of a dystopian film. Patients locked up in the Jose Pearson TB Hospital, surrounded by three razor wire-topped fences. A prison for the sick; no release dates unless the patient is in a coffin. After a number of patients staged an escape in a desperate bid to spend some time with their families after sometimes years of enforced isolation, they were arrested and the number of guards quadrupled. Quotes from health department spokespeople use the words 'hunted down', as if their individual rights to freedom were invalidated the moment they contracted XDR-TB.

I asked Phumeza whether she experienced any isolation during her treatment. She told me, 'I've never been in prison before, but while I was in hospital I was in prison. Everything is calculated, you are told what to do and when to do it. Family members have limited visiting hours. One is isolated from the outside world.' She spent nine months in this hospital before being moved to a hospice where visiting was easier and weekend passes were granted. In her blog, she talks fondly of the support and care she received at the Lizo Nobanda clinic. She is less complimentary of the Brooklyn Chest Hospital, saying: 'In big hospitals you will be left alone in a room and you will not be allowed to chat with your fellow patients. You are not even allowed to watch television. I HATE THEM FOR THAT.' This isolation and its negative effects on a person are unfortunately a common theme in reports from drug-resistant TB patients around the world. The problem is that a patient is considered infectious as long as there are bacteria in their sputum, yet once the drugs have started to take effect they are not sick enough to really need hospital care. So they're left sitting in a tiny room without much to entertain them. Phumeza never believed that this would happen to her: a healthy, HIV-negative student. I've read numerous reports from other patients who found it equally hard to accept that they were infected not only with TB but with a drug-resistant form. I think that's one of the big issues with TB – it is still an 'us and them' disease that happens to other people in other parts of the world. This isn't the case, though.

While researching this book, I read a quote about how XDR-TB is just a plane ride away. So it's fitting that the first US case I heard of was in a US lawyer who caused a worldwide debate when he flew to Europe for his wedding despite knowing that he was suffering from a drug-resistant infection. While he was abroad, doctors back home upgraded Andrew Speaker's infection from MDR to XDR-TB and everything got a bit crazy. Unwilling to be quarantined in Italy where he believed he might die, Speaker boarded a plane to Canada and drove across the border to the US, where he was ordered into medical isolation by the Centers for Disease Control and Prevention (CDC). His flight itinerary lists him as potentially exposing over 1,000 other passengers during his continent-hopping endeavours, but as far as I know no one else caught

TB from him (it's not that easy for the disease to spread during even a seven-hour flight, and Speaker wasn't obviously sick). Speaker's infection was downgraded again to MDR-TB and he underwent surgery to remove part of his infected lung. A happy ending for all, it seems, aside from the questions raised over when a patient becomes a criminal and an individual's rights to refuse treatment even when it may place others at risk.

I don't want to frame this chapter as 'you should care about TB because it might happen to you', as it seems like unnecessary fear-mongering. Cases of TB in Westerners like Speaker are relatively rare (that's why they end up in the newspapers). However, when they do occur, the advanced medical system in countries such as the US springs into action and the patient gets the care they need. Low- and middle-income countries, in comparison, can struggle under the huge burden of cases. People like Phumeza, through no fault of their own, don't always have access to the best diagnostics or to the few drugs capable of treating their infection. So instead of 'it could happen to you', I'm going to say 'it can happen to people *like* you'. Normal, everyday people who don't expect to catch XDR-TB and who are just as horrified to find themselves locked away from their families as you would be. Who, like many people in the Western world, associate this disease with poverty or HIV infection or the homeless, and think 'TB won't happen to me'. If their infection turns out to be a drug-resistant one, the unfairness of the situation is magnified. Drug resistance isn't just a problem for people who don't take their medication. More and more individuals are picking up MDR or even XDR-TB as their first infection; others are developing it through a failure of the healthcare system to correctly manage their treatment.

In Phumeza's case, it was a lack of diagnosis that nearly cost her her life. As a result of her experience, she co-authored a Doctors Without Borders drug-resistant TB petition calling for universal access to drug-resistant TB diagnosis and treatment. The 'Test Me, Treat Me' manifesto focuses on both the need for shorter, safer and more effective treatments and the issue of how drug resistance testing is still not routinely

carried out in many settings. I personally see this lack of
diagnosis as one of the biggest barriers to TB control. The last
few years have seen the introduction of two new TB drugs –
bedaquiline and delamanid. But if these drugs are simply added
to an already failing standard regimen, then resistance is going
to emerge even faster than what we've seen with the older TB
therapeutics.

Currently, the global prevalence of MDR-TB is estimated
at 3.3 per cent for new cases of TB and 20.5 per cent in
previously treated patients. It's not a uniform distribution
from country to country, though. In South Africa, just 1.8
per cent of new cases and 6.7 per cent of previously treated
cases are thought to be drug resistant. At the other end of the
scale, these numbers rise to 34 per cent and 69 per cent,
respectively, in Belarus. In many former members of the
Soviet Union, it's not uncommon to pick up MDR-TB from
the get-go. Here, drug-resistance is as far from an individual
problem as it can get. What worries me the most, however, is
the scale of the detection gap. The numbers I've quoted above
are estimates, not actual notified – and treated – cases. The
biggest gap seems to be in China, where only 11 per cent of
drug-resistant TB is detected. The problem is compounded
by less than 50 per cent of diagnosed drug-resistant TB
patients in China being started on second-line treatment,
although this number averages out at a more respectable 90
per cent-plus on a global scale. Treatment success is another
story. Less than half of drug-resistant patients make it through
to a positive outcome thanks to high death rates and people
slipping through the healthcare system's fingers and vanishing.
Clearly, urgent changes are needed to address these shameful
statistics. One day, those not-so-blue skies could well come
crashing down again, and we'll find ourselves in a situation
where a far bigger proportion of global TB cases are effectively
untreatable. Drug-resistant TB isn't someone else's problem;
it's already closer than you think.

CHAPTER ELEVEN

A Barometer of Inequality

'Parts of London have higher rates of TB than Iraq and Rwanda,' cry the headlines. The entire British press appears to have read a 2015 report from the London Assembly, and all of them have picked out this pithy comparison. What's not to love about it? I mean, Rwanda is in sub-Saharan Africa and Iraq is, um, in that general direction. TB's obviously going to be over there with, um, *them*. And now it's here too! 'It's the migrants', screech the online comment sections. Mr Angry thinks it's 'an obvious conclusion of our government welcoming the scum of the earth into our country'. Cerberus wants to 'boot these diseased people back to were [*sic*] they came from'. Steve SW has 'heard that napalm is a fantastic disinfectant'. Welcome to twenty-first-century Britain, where all the frothing-at-the-mouth racists bellow louder than the rest of us and consistently miss the point.

Let's take a closer look at that well-publicised comparison. I get that it was intended to show the scale of the problem in London. But Iraq and Rwanda are doing pretty well in terms of TB compared to lots of places in the world. On a TB heat map of the African continent, Rwanda is a light-yellow spot in a sea of dark reds. An oasis of success in TB control, where the incidence is 85 cases per 100,000 people (85/100,000), down from 118/100,000 in 2006. Iraq has an incidence of 67/100,000 – higher than its neighbours in the Persian Gulf, but that's hardly surprising. To put things in perspective, Rwanda's neighbour, the Democratic Republic of the Congo, has a 532/100,000 incidence, and South Africa is up at 834/100,000. In the TB league tables, Rwanda and Iraq don't get a look in. They barely make the top 100 countries, in fact. Both have some way to go before they join Western Europe with its sub-15/100,000 levels, but they're hardly sinking into an abyss of consumption and death. So I personally find the use of these two countries as a point of reference slightly confusing. It feels a little like playing on our built-in prejudices about African or Arab countries being hotbeds of disease and misery. Along the lines of: 'If things are as bad for London as Rwanda, then things must be really horrific!'

I'm not for a moment suggesting that the authors of the London Assembly report or the journalists who wrote about it were being deliberately racist. But my problem with this kind of reporting is that it reinforces the stereotype that TB is a disease of foreigners. An 'us and them' infection that, in the wrong hands, is taken to mean it's somehow the fault of those who've contracted it. Yes, a lot of the TB in London is in non-UK-born populations, but the home-grown problem is still there. And, yes, quite a bit of TB is thought to be reactivation of latent infections picked up in a person's country of origin, but the role of transmission on English soil is also not to be sniffed at. And, yes, I'm not denying that TB is pretty damn expensive to treat, adding to the troubles of our over-burdened health service. But I really wish the argument didn't have to descend into a debate over whether someone coming from abroad deserves to survive a curable disease. Because all this bitching about the re-importation of a long-gone disease is nonsense. It detracts from the real problem, which is that TB never went away in the UK. It isn't simply a disease of migrants. A lot of the time, it's a disease of inequality, poverty and a healthcare system that isn't doing enough to diagnose and treat TB in the most at-risk populations. If you want to assign blame, then the problem isn't migrants coming into 'our country' and spreading disease; the problem is 'our country' providing the perfect conditions for both non-UK-born and British people to succumb to an infection that very rarely affects those with healthy immune systems. Unfortunately, I'm not sure that's a distinction certain swaths of the population will agree with.

Perhaps a better way of thinking about TB in London is in comparison to the rest of the UK. Based on the stats used in the London Assembly report, London accounts for around 40 per cent of TB cases in England, with one-third of London boroughs exceeding the WHO's 40/100,000 threshold of 'high-incidence'. In 2014, there were more than 2,500 new cases of TB in London. Some of the worst affected districts included Brent (83/100,000), Ealing (65/100,000), Harrow (60/100,000), Hounslow (64/100,000) and Newham (100/100,000). Overall, the rate for England in 2014 was 12/100,000, but in London it came in at 30/100,000. It's clear that in certain pockets in London – and in other urban centres in the country – TB is a disproportionate problem.

London is my favourite place in the world. I love almost everything about it, from sunburned picnics in the park to early-hours dirty kebabs; the street art to the view from Waterloo Bridge at night; the crowded markets to trendy coffee shops. Now that I've spent a sufficient amount of time living outside the capital, I've even started to almost miss the cyclists who never stop for red lights, summer on the London Underground and queuing for two hours to buy a £6 pint. Almost. Its diversity is a big part of what makes London great, yet some of the poorest communities in London are immigrant and refugee populations. The borough of Newham, for example, has the largest number of foreign-born residents in London. It is also one of the most deprived areas in the country, along with its neighbours Hackney and Tower Hamlets.

So, teasing out the overlapping contributions of migration and poverty to TB rates isn't simple. Among Newham's problems are high levels of homelessness and poor housing conditions, including overcrowding in around 25 per cent of households. I was reading about a rented three-bedroom house in the area that, when raided by police, was found to house 26 people including a three-year-old child. In the tiny basement room, seven people slept on metal bunkbeds and a double bed. Perfect conditions for the spread of TB among people weakened by poor diets. One study put the rates of TB in the most deprived areas of England at seven times that of the least deprived. I talked to Catherine Cosgrove of the TB referral unit at St George's Hospital, London. She told me how some of the biggest challenges in treating TB have little to do with the disease and more to do with ensuring that a patient is able to follow through with the long treatment and regular hospital appointments. Many of her patients are working in jobs that, if you don't turn up, you don't get paid. Given the choice between feeding your family and seeing to that irritating cough that's not going away, I can understand why it's difficult to get people to the clinic in the first place, never mind keep them on treatment. 'A lot of these people have questionable immigration rights, and even if they're working legally they might not have full rights,' she told me. 'The minute you say they can't go to work, the dodgy one bedroom that they were renting from who knows who gets lost, so then you've got someone who is homeless with TB, and that's a huge problem.' Treating TB becomes a team effort that can involve wrangling accommodation for struggling patients and liaising with TB charities that pick up the slack where stretched NHS budgets can't reach.

There's an oft-quoted statistic that three-quarters of notified TB cases in England are in non-UK-born people, with those coming from India, Pakistan and Somalia making up the majority. In around 50 per cent of these cases, the patient develops active TB within five years of entering the country. These cases are likely reactivations of latent infections acquired in the patient's country of origin. What is it about coming to the UK that triggers someone's previously asymptomatic infection to turn rogue? One theory is that it is down to vitamin D deficiency. Vitamin D plays a role in helping the immune system kill invading pathogens, and a number of studies have shown that low levels can increase a person's risk of active TB. Move to the UK, where a combination of poor diet and Britain's notorious lack of sunlight results in falling vitamin D levels: reactivation TB! A recent paper from Ajit Lalvani's lab in London also suggests that vitamin D deficiency is a risk factor for the spread of TB from the lungs to other parts of the body. I'd always thought that TB was predominantly a disease of the lungs, with extra-pulmonary cases being confined to children of those with an HIV infection. Turns out I was wrong. In the UK, around 50 per cent of cases in adults are in non-lung parts of the body – the lymph nodes, bones and joints, the spine, the gastrointestinal tract, the brain. One of the most immediate problems here is that it makes diagnosing TB in these patients quite difficult as their symptoms can mimic a great number of other diseases. Their path to the TB referral unit can involve various departments before they finally find themselves with a diagnosis of TB. 'Most commonly we'll see lymph nodes,' Catherine Cosgrove told me. 'A lot of them will have come through the cancer pathways. I always try to tell people what marvellous news it is.' She paused to laugh before continuing, 'You thought you had cancer, but it's great news, it's only TB.'

How great the news is is a matter for debate. Professor Onn Min Kon, the lead clinician for the TB service at Imperial College Healthcare NHS Trust, told me: 'Incredible as it may sound, even with our first line of drugs for TB there's tension about getting drug supply internationally. It's a problem. Sometimes we run out of standard drugs. You'd be surprised at how many times you get these national alert things saying this drug is about to run out of supply.' A 2011 paper revealed that more than half of surveyed NHS TB treatment centres were struggling to obtain anti-TB drugs. A huge 27 per cent had to interrupt treatment at least once; 19 per cent resorted to altering the regimens. When it came to second-line drugs, 36 per cent reported difficulties in obtaining these medications. The biggest problem was in providing treatment for children, who require liquid formulations at lower doses than adults. In 26 per cent of the treatment centres that answered the questionnaire, children were being given unlicensed, variable-strength liquids and locally prepared suspensions.

Drug availability is just one of the problems associated with treating a disease that occurs relatively rarely within the population as a whole. The biggest issues are in diagnosing people in the first place. The average UK TB patient is a young migrant who has recently entered the country, possibly living in unsecured accommodation and working on a zero-hours contract, which means they are already struggling to put food on the table. They're probably too busy to see a GP, and if they do, there's no guarantee that the doctor will have come across enough TB to spot it straight away. A cough, weight loss and night sweats? Well that's not too hard to diagnose. Weight loss alone, though? Or night sweats with none of the other symptoms? In these patients, there are many, many diseases that may be higher up the list before someone considers TB. Even after diagnosis, there's the stigma associated with TB that stops people from wanting to let their contacts know that they need to get checked. 'Anyone can catch TB, that's the reality,' says Prof. Kon (as he introduces himself). Except not everyone realises TB is an ongoing health issue that can affect young, seemingly healthy individuals.

While anyone can, in theory, catch TB, it was levels of TB in migrant populations that were responsible for the big increase in England's TB rates between 1998 and 2008. Overall, rates went up by 40 per cent (corresponding to a 5 per cent drop among British-born people but a 94 per cent increase in cases in foreign-born persons). The trend has now reversed, solely down to improvements in the non-UK-born rates. This is, at best guess, down to a decrease in migrants arriving from high-burden countries. Good old-fashioned home-grown TB? Well, that's remained static for years, indicating that current control measures are doing little to improve things for this population of TB patients. I talked to Alistair Story, founder of London's Find&Treat service. At the end of the twentieth century, TB levels in London were on the rise and the existing control strategies were failing to reach the hardest-hit populations. Al's approach was along the lines of 'if people with TB symptoms won't go to the clinic then the clinic will go to them'. The team borrowed a van and parked up by a hostel next to a prison. Using a portable X-ray machine, they screened the residents of the hostel for TB. The success of this project proved the team's point about needing to get out into the community to find new cases of the disease. By 2005, they had their own vehicle, and quickly worked out that their efforts were best spent looking for TB in rough sleepers, those with substance-abuse problems and those who'd spent time in prison. Note: not migrants.

This reasoning was explained in a 2007 paper from Al's research team looking at the rates of TB in London among different groups. Recent migrants came in at 148 cases per 100,000 people. It's a rate 10-fold higher than that for England as a whole, so obviously not great. But when you look at problem drug users, they have a rate of 354/100,000. For people who've spent time in prison, it's 208/100,000. Those living in hostels or on the street are right up at 788/100,000. Overall numbers of cases might have been less for the homeless when compared to recent migrants (due to differences in population sizes), but the TB problem was clearly far more out of control in those living on the streets. Based on these numbers, actively screening 100 migrants for TB would find just one case. The homeless have eight times the incidence, and the added problem that they're less likely to pop along to the GP's to get a chronic cough checked out. Since 2007, the number of rough sleepers in London has more than doubled, meaning there is a constant stream of susceptible individuals to keep Al's team incredibly busy. He told me how they screen around 10,000 people a year and, since starting the project, have seen a steady detection rate of around 200–250 cases per 100,000. He describes it as a tap that can't be turned off. No matter how many people he finds and refers for treatment, more and more contract TB as a result of their chaotic lifestyles and general poor health.

The difficulties of working with these populations become evident when Al explains that, at first, 53 per cent of those he diagnosed were lost to follow-up. Somewhere between diagnosis and treatment, they just vanished from the medical system's radar. Hospital services were simply not set up to manage these complicated cases, so in response Al's team grew and diversified to include both social and medical care, with an emphasis on the social side of things. It's an integrated approach that focuses on the person and not the pathogen, as they put it. They don't just test for TB, but help patients to address their chaotic lives, improve their mental health and work on all the other illnesses common among these vulnerable populations, as well as the underlying determinants of disease. One of their big successes is raising TB treatment completion rates to match those of the general population, yet new cases keep on emerging. Unfortunately, the Find&Treat team struggle to get the funding they need because they don't fit into any local authority's remit. On top of this, homeless people simply aren't very fashionable at the moment (were they ever?). There's also not enough people living on the streets with TB to earn them a starring role in England's TB control strategies.

In 2015, Public Health England named TB as one of its seven priorities in greatest need of improvement. The others, if you're interested, were: smoking, obesity, alcohol, dementia, antimicrobial resistance and ensuring children have the best start in life. The same year, Public Health England published its Collaborative Tuberculosis Strategy for England, in which it outlined its reasoning for focusing on TB. Basically, if you draw a graph of TB cases, England's has been pretty much a straight line for some time. The US, in comparison, has followed a nice downward trajectory for the last 10 years. So something is obviously going wrong in the UK. Eradicating TB in England will mean finding a way to target the specific populations where TB remains a problem – where the clinical approach of 'diagnosis, drugs, discharge' isn't good enough. Migrants, it seems, are the first in the firing line. NHS England have invested £10 million into the systematic implementation of new-entrant TB screening, illustrating just how important they believe this to be in their overall strategy. Is it a good strategy? It's a start. I should say that one of the likely reasons why the US has been more successful at reducing TB levels is its aggressive approach to treating latent infections. I don't have a problem with this part. It's the specific focus on migrants that worries me, especially as there's not much evidence to suggest that it will have any discernible impact on the public health of the resident population.

Migrant screening for disease has a long history. In the early twentieth century, the huge surge of mainly Europeans arriving in the Americas brought with them diseases including

TB. I saw some pictures of 1917 arrivals at Ellis Island instructed to climb some particularly steep steps carrying their heavy luggage. Anyone coughing at the top was examined for TB; everyone smiling and looking happy was evaluated for mental illness and syphilis. Presumably, the best expression to wear would be one of boredom to avoid being shipped straight back to where you'd come from. Until fairly recently, the UK also screened new arrivals coming into Heathrow or Gatwick airports for TB. However, a 2016 study revealed that of over 200,000 people screened, only 59 diagnoses of TB were made over a period of 15 months. Chest X-rays are only capable of detecting active pulmonary TB and not extra-pulmonary disease or latent infection. Plus, not everyone enters the UK through these two airports. As cost-effective ways of stopping the importation of TB go, it's not got much on those TB-versus-syphilis stairs.

The UK now operates a full pre-entry system in which those from high-incidence regions applying for a long-term visa are screened in their country of origin. Should they be found to have TB, they are required to receive treatment before reapplying. Does it work? Well, one early study of 476,455 visa applicants found a culture-confirmed TB rate of 92/100,000 – that's a fair few cases prevented from being brought into the UK. Only, looking for active TB is a one-shot scenario. It misses the vast majority of TB because it fails to identify most cases of latent disease. This is where that £10 million comes in. A new entrant arrives in the UK and registers with a GP. They're tested for latent TB infection (LTBI) and, if necessary, treated. No reactivation TB; everyone is happy. Only, it wouldn't be cost-effective to screen everyone. Few coming in from, say, the US or Australia are going to have TB. However, focusing on only the highest-incidence countries of origin would miss a lot of cases. India, for example, has an incidence of only 167/100,000 (compared to South Africa's 834/100,000), yet people from India accounted for 16 per cent of visas per year, or 86,117 immigrants in total. Ajit Lalvani led a 2011 study revealing that nearly 90 per cent of all the LTBI cases being imported into the UK are in patients from the Indian subcontinent and sub-Saharan Africa. A cost-effectiveness analysis suggested that screening individuals from countries with an incidence of over 250/100,000 would avert 12 cases of TB in every 10,000 people over the course of 20 years (that's around 12,000 cases in total, presuming UK migration averages 500,000 newcomers a year for the next two decades). The cost would be £17,956 per case. Dropping the cut-off to 150/100,000 – to include the Indian subcontinent – would prevent an additional 29 cases per 10,000 immigrants, at a cost of £20,818.80 per case. This is the cut-off that the Collaborative Tuberculosis Strategy for England has in fact chosen. Something that stood out to me in Lalvani's study, however, was that they estimated there would be 95 cases of active TB per 10,000 people over 20 years. That's 95,000 cases, of which under a half would be averted by screening.

Mike Mandelbaum of TB Alert directed me to a paper that raises another problem. Of 252,368 new entrants to the UK whose 2012 entry was documented at Heathrow and Gatwick, how many do you think actually registered with a GP? A grand total of 32.5 per cent. Sure, a lot of entrants are most likely outwardly healthy young people in no need of visiting a GP – my partner didn't see a single doctor between the ages of 18 and 35, until a freak hoovering accident reminded him that he was not immortal. But many of those lost to the system are actually those with the highest health needs. Cultural and linguistic factors are proposed barriers to registration, as are concerns among migrants over their immigration status. A retrospective study of TB in foreign-born people registered with GPs in Leicestershire calculated that LTBI screening could have prevented 60 per cent of active TB cases in this population. Given that only a third of new entrants are registering in the first place, surely this suggests that screening will catch only 20 per cent of cases in total? The £10 million committed by NHS England towards targeting LTBI in new entrants seems like a big price tag for something that, while having a big impact on TB, isn't going to get close to universal coverage.

Is the focus on migrants a politically motivated decision? An attempt to address one well-publicised problem rather than spreading the money more thinly to ineffectively combat all the various factors involved? According to some of the scientists I interviewed for this chapter, the best outcome of the LTBI screening approach would be continued anti-TB momentum that translates into better all-round TB policies. At worst, though, risk-based stratification of TB testing could end up diverting resources away from the areas that really need improvement, such as ensuring that people complete their treatment. It's a scary thought that when it comes to diagnostic delays and treatment completion rates, the UK is lagging behind many high-burden countries. A 2013 study in the Thames Valley found that the median overall delay in diagnosing TB was 73 days from onset of symptoms to the start of treatment. That's over two months during which time a patient could be infecting others or becoming increasingly sick. The median health-service delay alone – that's all the faffing about between a patient first seeing a doctor and getting an actual diagnosis – was 39 days. GPs have a lot on their plates, but they should be considering TB whenever they see an unwell, fatigued patient from a high-risk group, especially one who is losing weight and feverous or has been coughing for more than three weeks. Less 'all right let's do some general blood tests and why don't you try to get a bit more exercise as that can improve your sleep'. More 'I'm going to refer you to the TB clinic but don't worry as it's a treatable disease especially if caught early'. Education for GPs is within the remit of the only charity supporting UK TB patients: TB Alert. Their 'The Truth About TB' website is an interesting read, especially if you happen to belong to the most at-risk populations. You probably don't, though, and that's part of the problem.

Amy McConville also doesn't look like the traditional TB victim. I first came across Amy via a newspaper clipping and two mischievous co-workers who, while I was at lunch, sellotaped the headline 'TB has ruined my life but I'm fighting back' over my desk. It was cut out from an article on a

19-year-old white British student who found herself struck down by TB. The press jumped on her because her story was so unexpected. If it could happen to her then none of us are safe! But while patients like Amy make the news, in reality they represent the end of the line for *M. tuberculosis*. Transmission of TB is like the spread of a fire. First you need fuel: human hosts capable of supporting the infection. Some people are the equivalent of a damp log, while others – those with poor health and compromised immune systems – are dry grass. Amy's friends and acquaintances weren't homeless; they didn't have drug- and alcohol-abuse problems. So already her TB infection would have been hard-pushed to find a new host. You also need oxygen for a fire to burn. My analogy's oxygen is the right combination of conditions for the infection to spread. Again, Amy wasn't spending her time in overcrowded prison cells, homeless shelters or drug dens, nor was she living with 25 others in a small three-bedroom house. Finally, you need a spark – or an infectious TB patient. Some cough up a lot of bacteria, others don't. It's the difference between a tiny ember and a burning branch. Let's imagine a really poorly person with a range of co-morbidities and advanced, untreated TB. They're coughing, coughing, coughing, and a lot of equally unhealthy people are sharing the same crowded, infectious air. My point here is that while transmission won't occur without this patient, they're only part of the problem.

I was reading a fascinating book chapter by sociologist and historian David Barnes on the problem of searching for the person responsible for igniting an outbreak of TB – the index case or, going with Barnes' terminology, Patient Zero. He articulates my own feelings on the ethics of public health interventions for TB better than I can, so I'll point anyone interested to a 2010 book called *Tuberculosis Then and Now: Perspectives on the History of an Infectious Disease*. My own discomfort with the whole idea of Patient Zero ties in with all the panic over migrants 'bringing TB into the country'. It strays dangerously close to laying blame on individuals rather than focusing on the social determinants of disease transmission. For what remains of this chapter, I want to talk about the epidemiological tracking of TB epidemics. I love these studies. They can teach us so much about how disease spreads and the risk factors associated with transmission. They are basically like little scientific puzzles in which an outbreak can be traced back to its starting point or individuals identified as 'super-spreaders' who, through a combination of coughing up lots of bacteria and spending time in close proximity to other susceptible hosts, contribute hugely to the spread of the disease. We totally need to find those people, I hear you say! They are the ones whom we should be targeting to halt the spread of an infection. That's true, I'll concede. Yet the cynic in me worries about how public-health policies or public opinion might, in the future, interpret this sort of information. Just look at Andrew Speaker. According to Barnes, media outlets called for him to be charged with attempted murder for putting all those fellow airline passengers at risk. And during the 2003 SARS epidemic, China apparently threatened to execute anyone spreading the virus. Extreme examples, but TB control treads a fine line between ensuring someone receives treatment (through supervised pill-swallowing, for example) and criminalising patients, as seen in the prison-hospitals for XDR-TB patients. Surely not in the UK, you say? So what about nurse Pauline Cafferkey, who may be struck off the medical register as a result of accidentally putting others at risk of contracting Ebola? My pessimistic attitude is that, even in the UK, we are just one health scare – Ebola, bird flu, drug-resistant TB – away from a George Orwell novel.

Can we, for a moment, pretend that the world we live in is a more utopian than dystopian place, where groups of individuals (migrants, the homeless, prisoners, drug users) are not stereotyped and scapegoated? Let's follow a TB outbreak and see how epidemiological tracking exercises can target medical interventions – non-judgementally – to those in need. In January 2000, microbiologists at a North London hospital noticed something strange. While they were used to seeing the odd case of TB, four had popped up in the same week, all in young men from the local community. All of the isolates had the same resistance to isoniazid, suggestive of an outbreak. Dun-dun-duuun! I like to imagine that this next part was like in a thriller where a detective notices similarities between two murders and goes through old unsolved case files to discover there's a serial killer at large. He then probably has a swig from the bottle of whiskey stashed in his desk ... while, from my own experience, microbiologists are more partial to tea. Anyway, our plucky microbiologists looked back at all their isoniazid-resistance strains isolated between 1995 and 2000, using a technique called IS61100 restriction fragment length polymorphism (RFLP) analysis. Back in 2000, this was one of the key methods used for fingerprinting strains of *M. tuberculosis*. The technique uses a restriction enzyme – a biological pair of scissors that cuts DNA at very specific sequences – to assess how related two isolates are. Same fingerprint of chopped up DNA equals same DNA sequence, indicative of transmission rather than the independent emergence of drug resistance. RFLP analysis revealed that the four isoniazid-resistant isolates were actually part of a much larger outbreak focused around North London. Everything had started with one individual who had likely brought his strain with him from Nigeria.

A Public Health Laboratory Service (PHLS)/Communicable Disease Surveillance Centre (CDSC) multi-district Incident Control Committee was formed in June 2000. You know things are serious when there are more than 10 words in the name of your committee. By the end of 2001, 70 cases in total had been identified, either retrospectively by comparing RFLP patterns to older isolates or by targeted case detection, most commonly contact tracing. Someone with active TB is interviewed to identify anyone who's spent a lot of time with them, such as family members and social or work contacts. Those identified can then be screened for TB, and if any active cases are discovered, the whole process starts again with the new patient at the centre. By identifying new cases of TB early, the chain of transmission can be broken before anyone else is infected. It's a method used in the control of many infectious diseases, but which has particular importance to those which are hard to treat or highly fatal (Ebola, for example, relies solely on stopping transmission, as there's no real treatment once you have the disease). As TB drug resistance becomes more and more of a problem, the importance of controlling outbreaks increases. No one wants a resistant strain to take hold in a population. So the North London isoniazid cases were a particular concern, especially as a 2004 paper identified a worryingly high 11 per cent transmission rate between close contacts, compared to 0.7 to 2 per cent in other documented outbreaks.

What this outbreak demonstrates nicely is that contact tracing doesn't always work too well in hard-to-reach populations. You can imagine the spread of TB as intercepting concentric circles. Our index case was socially linked to 18 new patients, almost half of whom were recreational drug users. Each of the 18 then becomes the centre of their own circle as a potential new starting point for additional clusters of disease. Four of the 18 could be linked to a further 10 cases by virtue of everyone being held in the same prison. One of the 10 then took his TB to an intravenous drug-user squat where between 7 and 12 people also contracted the disease. As the circles spread outwards, the outbreak found its way to two healthcare workers, a brother and sister from North London and a couple of businessmen, among others.

But these people, like Amy McConville, are more likely than not to be the end of the line for TB. By 2004, 30 of the 70 initial cases had completed treatment; three had died (although only one from TB), four had disappeared into the depths of London/Nigeria and the rest were continuing treatment. At least 17 of the cases were found to be extremely unreliable when it came to taking their medication and turning up for appointments. A further 10 raised suspicions about whether they were complying with their treatment regimens. As I said earlier, I'm not into assigning blame to individuals. Many of the problems in working with these patients appear to stem from their circumstances rather than any deliberate wish to put their own health, and that of others, at risk. As with many medical problems in the UK, there was a heap of red tape involved in getting patients the medication they needed. Then there was the issue of patients needing to travel to the clinic on a daily basis to receive treatment. In some cases, TB nurses had to pay out of their own pockets to ensure patients could afford the bus fare. Come on, Public Health England, a £1.50 bus ticket should not be determining whether an outbreak of TB is brought under control or not. What all of this means is that it's not really surprising that, 10 years on, the outbreak was still happily simmering away. By 2013, the cases numbered more than 500.

This was a treatable strain of TB. Imagine what would happen were highly drug-resistant TB to start transmitting so easily among vulnerable London populations. It's an eventuality that's closer than you might like to imagine. In April 2013, an individual with a nine-month history of cough turned up at a London hospital and was diagnosed as being infected with XDR-TB. Control measures kicked in and this index case was isolated from the rest of humanity. In all, he ended up spending seven months in hospital due to his whole left lung being riddled with TB and the right not faring particularly well either. Contact tracing revealed 33 named contacts who'd potentially been exposed to TB; two of these had contracted probable active disease in addition to 12 with latent infections. Looking back retrospectively on the outbreak, however, revealed that there had been three additional active TB cases, one of whom had died. There's an interesting section in the paper describing the outbreak that explains how the daily diary of Case 5's activities revealed they were spending quite a lot of time at 'a local amenity'. 'Further questioning of the index case did not confirm attendance at the amenity, however third party information suggested that case 1 had spent time there whilst infectious 2 years earlier.'

I'd love to know what the amenity was – I'm guessing it was something that the index case was loath to admit to. But, drug den or My Little Pony appreciation club, this raises a big issue with contact-tracing exercises. You're basing everything on a patient being accurate and honest in the interview.

This is where whole genome sequencing comes in. Again. This book is becoming a '101 uses for whole genome sequencing' guide. This time round, it's the potential of whole genome sequencing to follow outbreaks in real time that is getting people excited. I don't know if you've ever read a scientific paper, but on the whole they generally don't contain too much wild speculation that can't be backed up by data. Except for the last few sentences of the conclusions. Here, scientists get to profess their hopes for the future of their work without fully addressing whether it's feasible in reality. It's in these short paragraphs that I keep reading how whole genome sequencing has the *potential* to allow us to monitor the transmission and evolution of pathogens in real time. It sounds great, but is it really something that's on the immediate horizon? I called up Lukas Fenner in Switzerland to find out. Lukas recently published a paper demonstrating how targeted whole genome sequencing can be used to create a network of transmission. It was an ambitious study investigating 1,642 TB cases spanning two decades, the majority of whom were infected during a known peak of heroin abuse in Switzerland between 1991 and 1995. After narrowing the isolates down to those that stood at the centre of clusters, the authors used whole genome sequencing to look for a genetic signature. This signature could then be used to differentiate between strains using a simple PCR test. They combined the results with social contact information to retrace transmission of the strain at high resolution.

The overall network of transmission looked like three overlapping fireworks, with super-spreaders at the centres of each burst, reaching outwards to infect large numbers of their contacts. In real time, the technique could have revealed transmission hotspots where additional patients might be discovered, found missing links in the transmission chain in need of treatment or even provided advance warning of the emergence of drug resistance had it occurred. This study, however, was retrospective, and Lukas is cautious about the viability of this technique to monitor outbreaks as they happen. The problem isn't whole genome sequencing technology or the cost – it's the huge amount of data generated. We're not there yet in terms of user-friendly software that can make sense of genomic data. Lucas mentioned how the UK in particular now has a fantastic set-up when it comes to whole genome sequencing of strains but, even so, he believes we are looking at data analysis times of a few weeks, not real time. For now, at least. If we do get there, I doubt it will be TB leading the charge. There simply isn't a huge need for such technologies in low-incidence countries where TB outbreaks are relatively rare. Instead, it's the spread of nosocomial infections such as MRSA or drug-resistant *Clostridium difficile* where we have the biggest problems, and I suspect it will be here that the technology really makes the leap from potential to reality. How these technologies will trickle down into the control of TB, especially in high-burden countries, remains to be seen.

In any case, the current technology isn't the biggest barrier to TB control. 'Sophisticated machines don't solve all the problems,' Lukas told me. He's heavily involved in TB research in Tanzania, and he believes that getting patients to the clinic in the first place is the real problem. 'Health-seeking behaviour, the social culture and context, the religion, gender issues. These are really hugely important factors that have somehow been forgotten.' While the WHO acknowledges this, there is a need to find ways to address it in the field. It's funny how, no matter where in the world you stand, TB control suffers from the same problems. What TB achieves in low-incidence countries is to shine a light on these social determinants of disease and the growing pockets of inequality in our major cities. I've focused only on London here, but TB remains a problem in many other urban centres. The United States, for example, has problems among the homeless, where the rate is around 10 times that of the general population.

While homeless individuals make up 6 per cent of US TB cases, they comprise 20 per cent of clusters. Not only this, but stats from the CDC indicate that 85 per cent of TB occurs in racial and ethnic minorities. Yet another example of how TB is a far bigger problem in those with greater obstacles to obtaining quality healthcare. The London Assembly report I talked about earlier referred to TB as a 'barometer of inequality'. It's a good description so I've stolen it for the title of this chapter.

I've talked about London here for the simple reason that the UK is doing worse when it comes to TB control than many other high-income countries such as the US. The rate in England is five times higher than in the US and, within two years, we will have more cases of TB than the whole of the US. A country with a population five times bigger than our own. The improvements required to control TB in England are much the same as in any low-incidence, high-resource country where TB is clinging on among the most disadvantaged populations and finding new footholds as patterns of migration change. The small world that we live in means that, until TB is brought under control in high-burden countries, it won't be possible to completely eradicate it from urban centres such as London. But we can certainly do a lot better, and my hope is that the push from Public Health England to target TB will make some real differences.

Reducing the pool of latently infected individuals, particularly those belonging to migrant populations who may not be entirely visible to the healthcare system, is only part of the solution. I see it as a backdrop against which a number of other interventions need to take place. First, we need to be able to actively find new cases among at-risk groups and treat them before they spread their disease further. This requires education among both patients and doctors so that the symptoms of TB are recognised quickly, in addition to the active case-finding work of people like Al Story. Second, we need to ensure that the healthcare system is set up to help patients comply with their treatment. Things like providing bus fares, encouraging migrants to register with GPs and reducing the red tape strangling England's NHS. Third, we should be keeping an eye out for drug resistance and treating it quickly and efficiently. This already happens, but improvements in diagnostic techniques will be a huge help (I'm coming back to diagnostics in the next chapter). I won't suggest that we 'cure' poverty and inequality as I'm trying to focus on interventions that are actually possible in today's world.

What all of this needs is political commitment and funding, and that seems to be the biggest sticking point for TB control right now. During the writing of this chapter, I've come to realise that a big part of the TB problem in England is political, and it makes me super-irritated. I spent 10 years of my life attempting to find new ways of fighting TB – as if a brand new drug would make everything better. Only TB is much more than a clinical problem, and as Al Story puts it, control interventions need to focus on the people, not the pathogen.

CHAPTER TWELVE

Ratting Out the Missing 3 Million

O n the nose, I'm getting a herbal bouquet of tarry roads on a hot summer's day, enhanced with a sickeningly sweet fruity aroma that catches at the back of my throat like nail polish. And what's that? A hint of stewed cabbage, perhaps, and top notes of over-fermented beer? Appearance-wise, we've got a creamy yellow colour with little clumps settling out at the bottom and a scummy ring at the surface. To be honest, it's kind of disgusting, but nothing out of the ordinary for a two-week-old culture of mycobacteria. We microbiologists have a thing about sniffing cultures. Not a face-in-flask sort of sniff; more of a gentle wafting to delicately carry those bacterial aromas to our noses. It's particularly handy when you work on a slow-growing species, where contamination with more prolific bugs is an occupational hazard. Smells yeasty or slightly faecal? Yeah, that needs to be killed and washed down the sink. However, working on TB rules out this useful albeit slightly strange technique. Airborne diseases of the fatal variety don't mesh well with laboratory-mediated inhalation. Then it happened! Nearly 10 years into my career, I started working on a hobbled version of *M. tuberculosis* that didn't need all those safety precautions. Word spread quickly among my vast network of researcher friends. Within the space of a month I'd had two, maybe three texts asking the same question: what does TB smell like? So next time I opened one of the spent flasks to kill its contents, I had a small waft. I'm not sure what I'd been expecting, to be honest. Something that reeked of Death. A stench that left you in no doubt that this was a Killer. A toxic cloud of billowing green gas that left students rolling on the floor, clutching their throats. Instead, I got an anticlimactic chemical smell with phenolic undertones and a vaguely unpleasant tang.

If my name was Mandy and I was a freakishly large giant African pouched rat, I'd be able to detect this characteristic smell with just one snuffle of a patient's sputum sample. These Rodents of Unusual Size, you see, have three times as many olfactory receptor genes as humans and will do almost anything for a delicious puree of banana and nut. Yes, they are literally willing to work for peanuts. Which is extremely handy for countries such as Tanzania that struggle to correctly diagnose every potential TB patient using existing techniques. That's what the area of TB diagnostics all comes down to: money … but we'll get back to that once we've talked R.O.U.S.

APOPO is a Belgian NGO based in Tanzania, famed for its work in training giant African pouched rats to sniff out landmines. One day, founder Bart Weetjens was considering another big problem in Tanzania – the lack of reliable TB diagnostics that can be used in resource-limited areas. Recalling historical reports suggesting that a consumptive patient's breath often had a tarry odour, and the fact that the word for TB in his native Dutch translates as tar, his thoughts went to his beloved rats. After all, if you can train a rodent to detect TNT, you're only one letter off TB. Weetjens' hunch was right – the giant African pouched rats soon demonstrated themselves to be adept at detecting TB's unique smellprint.

Mandy and her friends are part of a diagnostic safety net only required because the first-line diagnostic test for TB in Tanzania gets it wrong 20 to 60 per cent of the time. Around the world, diagnosing TB often relies on a technique called sputum smear microscopy. It involves staining a sample of a patient's coughed-up sputum and looking for bacteria under the microscope. It's quick, cheap and relatively simple. It's also relatively rubbish, missing many cases of TB. Sure, it catches patients who are producing a lot of bacteria and are consequently the most infectious individuals. But it doesn't do so well when it comes to diagnosing extra-pulmonary TB or disease in children or in those infected with HIV (all of whom don't tend to have many bacteria in their sputum). I read that in high TB settings where microscopy technicians are inundated by large numbers of samples, they can have less than 60 seconds to view one slide. Do you remember school science lessons, with all the 'stop shutting your left eye when you look through the microscope' and 'can someone help me focus please, all I can see is a gigantic worm-thing'? Granted, people who do this for a living are a lot better at using a microscope than the average 14-year-old. But 60 seconds still isn't long to whizz through a blue-stained smear of phlegm looking for a tiny purple blob. And the consequence of missing that blob is a patient being told they don't have TB. Then they will head home and not only get increasingly sick but also potentially infect others. This is where the HeroRATs (so called because they save lives) come in. Today, 28 clinics in Dar es Salaam, Tanzania, and 14 in Maputo, Mozambique, send any sputum samples determined to be negative for TB (based on microscopy) to be double-checked by the rats. It's quite a sad indictment of the TB diagnostics landscape that these backup rats are needed, but that's the world we live in.

I've watched videos of Mandy at work, and her levels of concentration and the speed at which she diagnoses TB are impressive. In a metal and glass box, she scuttles from left to right as a (human) technician sequentially opens small holes in the floor. A quick sniff and she's moved on to the next, and the next. Then the tap of her claws on metal ceases and she inserts her long nose so far into one hole that only her beady eyes and rounded ears are visible. She pauses for three seconds or so before a loud click sends her lunging for a food-filled syringe, where she feasts. Greedy Mandy has identified one of the samples as containing *M. tuberculosis* bacilli and her reward is a tasty treat. A HeroRAT can screen 40 samples in less than 7 minutes. The same work would take a human microscopy technician a full day to perform. Since the project started, the rats have screened nearly 400,000 samples for TB and identified close to 10,000 new cases that had been missed by the first-line microscopy. Sure, there are other diagnostic tools out there that, on paper, are more reliable than either sputum smear microscopy or rats. However, some of the newer technologies and modern techniques are only as good as the power supply in cities such as Maputo, *i.e.* not very good at all. With sometimes daily blackouts occurring without notice, often resulting from the tropical rains, it's not always possible to rely on anything that requires electricity. And that's in the cities. Mozambique is a country that, in rural areas, remains almost completely in the dark, with only around 1 per cent of the population having access to electricity. Rats don't particularly care about power or water supplies. They don't care about much other than eating and grooming, and, importantly, they can remove some of a setting's reliance on outside expertise and funding. It takes six to nine months to train a HeroRAT at a cost around $8,000, but that's a whole lot cheaper than some of the modern diagnostic techniques currently available. The rats are faster than any other technology, which helps make up for the fact that they don't get it right all the time. They're also more reliable than sputum smear microscopy, although that's not saying much. Since the start of the project, HeroRATs have resulted in a 45 per cent increase in the diagnosis of new TB cases.

One case was that of Lulu, who nearly died after her local clinic failed to diagnose her. The HeroRATs literally saved her life, and following her recovery she joined a volunteer group called MKUTA. They work in the clinics and local communities to educate people about TB and to help ensure speedy diagnosis and treatment of new cases. These volunteers are the ones who track you down if the HeroRATs find *M. tuberculosis* in your sputum sample. Without their help, many patients would be lost to follow-up. Thanks to the rats and the human volunteers, a whole lot of lives have been saved and new transmissions of TB prevented. But rats still aren't a perfect solution to what is a very complicated problem. For starters, they are quite prone to calling false positives in a large number of cases, meaning that each sample has to be confirmed using another method. I keep thinking about how, before I researched this book, I took our ability to diagnose TB for granted. Patient gets ill and goes to the doctor she discovers she has TB. It's after the diagnosis that all the difficult stuff begins. Except, no. Even today, close to one-third of new TB cases are not diagnosed. Three point six million cases, missed by the doctors or not making it to the clinics in the first place. APOPO's rats are just one of the intervention strategies designed to identify TB sufferers quickly and cheaply so they can receive the treatment they need. But how did we get into the position where rodents are a viable addition to all the science stuff in the first place?

In every country, from low- to high-incidence, the gold standard for TB diagnosis is culturing bacteria from a sputum sample. If you can grow *M. tuberculosis*, the patient has TB. No growth? They most probably don't have active disease (although there are exceptions). The problem is that *M. tuberculosis* is slooooooow to grow in the lab. While non-commercial culture methods can speed things up, the technology is yet to make its way into the field (for example, Catherine Cosgrove, whom I spoke to for Chapter 11, is involved in trials looking at improved culture mediums and techniques that may help). The best available right now are a number of commercial liquid-culture systems. The BACTEC MGIT is a widely adopted, automated method. Simply, a sample is mixed with growth media and incubated at 37 degrees Celsius (99 Fahrenheit) – lung temperature. A fluorescent sensor in the tube starts to glow under UV once growing bacteria have used up the available oxygen. No glow equals no growth; fluorescence equals a diagnosis of TB. If the sample is positive for TB, the same method can be used to check for drug resistance by re-culturing in the presence of TB drugs. This method can identify a positive sputum sample in a week or two, with another few weeks required to look for drug resistance if the patient is believed to be at risk. This isn't fast enough. Making someone wait a few weeks for a diagnosis (longer for the drug resistance results) might not seem like a lot. Except it's more than enough time for a patient to return home and disappear from the medical system's radar. The patient doesn't receive their results or any treatment; they become sicker and potentially infect others. We need to do better.

A couple of years ago, an expert panel got together and wrote their letter to Father Christmas asking for new TB diagnostics. The four top requests were: a triage test to rule out uninfected patients, a sputum-based test that can replace microscopy, a non-sputum-based test that can be used in HIV

patients or children and a rapid drug sensitivity test. The first
three of these would ideally be performed at the point of care,
quickly and by someone with limited training.

Let's say I'm a patient in a low-income country and I have
some very mild TB symptoms or am wondering if I might
have caught it from a friend. I'll go to the local community
clinic where someone will perform a quick test on my urine or
blood that will give an indication if I have TB or not, although
it probably won't be a clear-cut answer. If it comes up as high-
risk, I'll be referred to an actual clinic or hospital where I'll
provide a sputum sample (if I'm HIV-positive or a child, there'll
be an alternative that doesn't rely on sputum). This sample will
be used first for a definitive yes-or-no test, the results of which
will be returned before I leave the clinic so that I can be started
on treatment immediately. Ideally, the test will also give a good
indication whether my infection is likely to be drug-resistant
or not, although it won't provide a full resistance profile – that's
going to take the rapid drug-sensitivity test, which probably
won't be *that* rapid (but quicker than the current method). So
I'll toddle off back home where hopefully a community-based
TB programme will work with me to ensure I take all my
medications and provide me with the support I need to come
to terms with my diagnosis.

Once the results of my drug-sensitivity testing come in, I
might need to be switched over onto a more individualised
treatment regimen, but hopefully this occurs quickly before
my TB symptoms even start to really bother me or I spread
the disease to others. Everyone wins, except for *M. tuberculosis*,
who dies. In most cases, this happy outcome can be achieved
with the drugs that are currently available. That's an important
thing to remember – new drugs aren't the only answer to the
TB problem when we are currently unable to diagnose all
those people who have the disease in the first place.

When I first started learning about this area, the idea of
lots of different layers of diagnostic tests confused me. Why
can't they just stick it all into one test, I wondered? But then
I thought about all the other diseases that are diagnosed
using tests of increasing complexity, cost and expertise. So in
case you too are wondering about this, I'll explain how I see

it, using type II diabetes as an example. Your triage test here could be a waist circumference measurement. A big tummy obviously doesn't mean you actually have type II diabetes, but it's enough of a risk factor that you should get some additional tests. A UK diabetes charity, in fact, has a roadshow that pops up around the country to raise awareness of the condition and actively find people at risk of type II diabetes before they show serious symptoms, analogous to active case-finding in TB control. Should someone be found to be at high risk (their weight being just part of the picture), they'll head off to the doctors, urine sample in hand. Here, a quick dipstick test can reveal the presence of glucose in their urine, which is strongly suggestive of uncontrolled diabetes. If the test does suggest diabetes, they'll be sent for definitive blood tests to not only make a diagnosis but also determine the degree to which their condition has progressed, and will possibly be referred to an expert who can tailor their treatment as required. You wouldn't use a glucose-tolerance blood test – no eating for eight hours followed by several hours sitting in the hospital so the nurses can take blood samples – to screen for diabetes. Most people won't have the condition and it would be a waste of everyone's time and money. Same goes for TB. Hence the need for a triage test at the beginning of the diagnostic rollercoaster, progressing all the way up to expensive but informative drug-sensitivity testing that requires expert analysis.

So, where do we stand when it comes to a triage test for TB that can be performed in community settings? A shopping list of ideal requirements for the triage test would be: very fast, cheap, easy to perform, relatively accurate and not requiring a continuous supply of electricity or expensive kit. It needs to work in high temperatures and high humidity, and in dusty settings where regular complicated maintenance of equipment isn't routine. It will be performed by the first person to see a patient – that could be a healthcare practitioner at a local clinic or someone driving an active case-finding bus into town.

There's nothing on the near horizon, but biomarkers are the most likely answer here. Biomarker tests look for the pathogen's footprints or the imprint that it leaves on the immune system.

It's a bit like identifying a species of bear from the bite marks in your leg or the trail of bear poo and hair that it drops in its wake. In theory, a simple blood, urine, saliva or breath test could not only reveal whether someone has TB, but could also be used to predict patients at risk of treatment failure or reactivation of a latent infection. It wouldn't need someone to produce sputum, or even have bacteria floating around in their airways. Basically, you don't need to even see the bear in person to know where to set traps for it.

Biomarker tests can detect two flavours of biomarker – pathogen-derived (bear poo) and host-derived (bite marks). To date, the best example of the former is a urine dipstick test that detects lipoarabinomannan – a component of the TB cell wall. The test can be performed in 25 minutes with no greater resources than a collection tube and the dipstick. One problem: it's not very sensitive and misses a lot of cases. A recent clinical trial led by Keertan Dheda tested this new diagnostic in HIV-positive patients living in sub-Saharan Africa. These are the patients TB is most likely to kill and simultaneously the ones who are the most difficult to diagnose. With rapid treatment being key to saving the lives of seriously ill HIV patients, tests like Dheda's could make all the difference, especially considering it only costs around $2.66. His study used eight-week mortality as its readout – these patients were all very, very sick – and found that the simple test reduced mortality by 4 per cent. It's not much, but considering the 300,000 Africans dying of HIV-TB every year, it's enough to make a big impact. But is it really good enough to make it as a universal test for every TB patient? Not yet, no. That's the problem with everything biomarker-related in the TB world. It's all still at the point of 'this could potentially be brilliant!' but the research is at too early a stage to tip over into reality.

While writing this book, I kept hoping that someone would make a big biomarker breakthrough. Any day now, I kept telling myself, and then I'll have something I can actually write about. Claudia Denkinger is in charge of TB diagnostics at FIND, a global non-profit organisation catalysing the development and delivery of diagnostics for poverty-related diseases. She told me, 'I think in the next five years, we might be able to get to a biomarker-based triage test, and that would likely be a combination of several biomarkers, possibly combining pathogen and host biomarkers. But if you asked me to put a likelihood on that, I'd say it's maybe about 50, 60 per cent likely. I think there's a lot of risk, still.' When it comes to looking for a host signature indicating infection with *M. tuberculosis*, there's a lot of promising research out there. I love the idea of using host mRNA, microRNA, protein, metabolite or epigenetic profiling to indirectly detect the presence of disease and to provide a prognosis for an infected patient. But at the moment, most of the signatures obtained predominantly from transcriptomic profiling (RNA) are too complex to look for outside of the lab. Measuring the levels of even a dozen transcribed genes (rather than hundreds) is a tall order in the field. Those potential signatures approaching the target characteristics required for a real-life test are likely to lose performance when you start using them in the broader population, where they have to distinguish TB from a gazillion other diseases and conditions. It's also important to consider that the test has to work on patients who might not respond to infection in the same way as their next-door neighbour. 'I'm kind of cautiously optimistic but I haven't seen anything in the pipeline where I think "wow".' Claudia told me, diagnosing the whole field of research with a simplicity and accuracy that's so far lacking when it comes to biomarker tests.

Something that doesn't fully meet the triage test shopping list but is further advanced in terms of real-world usage is the humble chest X-ray. Chest X-rays have a big role to play in TB diagnosis – remember Al Story's van? X-rays are relatively cheap and easy to perform but, on the negative side, difficult to interpret. Depending on the operator and their current state of mind (have they had their morning coffee and a good night's sleep?), the same X-ray can be interpreted as revealing completely different things. It is, after all, a 2D representation of a complex 3D organ. Is that a malevolent shadow or just a shadow? What's that white blob? Oh yeah, the heart, silly me. Ideally, we need to take cranky, unreliable humans out of the equation altogether. A 2015 paper from Bram van Ginneken's lab in the Netherlands details a way of doing exactly this. It reminds me of the plot of

Terminator, except Skynet has been replaced by a less-threatening program called CAD4TB that is capable of automatically reading a chest X-ray. CAD4TB (other automated programs are available) takes one minute to process an X-ray and gives out a score for how not normal the lungs appear. Zero is 'well done, you're perfect' and 100 is 'something is terribly wrong here', although the program can't differentiate between all the different types of wrong that can afflict a pair of lungs.

Depending on where you place the cut-off of wrongness, fewer or more patients will be sent for further tests. Too low a cut-off and you'll spend a lot of time and money trying to culture sputum from everyone. Too high and you're going to save the pennies but miss the patients. Bram van Ginneken's team looked at 388 individuals with suspected TB who were subjected to chest radiography and sputum culture testing (the gold standard). Using a cut-off of 85 for the automated chest X-ray reading, they found that 40 per cent of patients would be sent for further testing with around a 50 per cent reduction in costs and an increase in patient throughput from 45 to 113 per day. This cut-off, however, would have led to 2 per cent of TB cases being missed, of which 1.5 per cent were HIV-positive (less easy to see TB on a chest X-ray). To improve on the sensitivity of the technique, the same authors more recently looked at using a combination of automated X-ray reading and an algorithm that takes into account clinical information such as symptoms and HIV status. When I talked to Claudia Denkinger, she told me that she was off to a WHO meeting on chest X-rays the following week, so clearly this is an area that has caught the attention of a lot of clever people. I always breathe a sigh of relief when I ask an expert a non-stupid question because, a lot of the time, the research that I find interesting can be quite 'niche'. Once, I pitched an idea for a news story to an editor and when he asked why it was worth covering, my only answer was 'because it has llamas in it'. Anyway, Claudia sounded cautiously optimistic about chest X-rays as a triage test but mentioned several negatives, including how they can't be used outside of hospital settings, simply because the electricity supply isn't there. I guess one way of looking at them is as an imperfect solution in an imperfect world. It would be nice if we didn't have to rely on them, but currently there aren't any better alternatives.

What everything I've discussed so far has in common, with the exception of the gold standard culture-based tests, is that it cannot tell you if a patient is infected with a drug-resistant strain of M. tuberculosis. Drug-sensitivity testing is a big deal and, in an ideal world, would be performed for every single TB patient. Remember Phumeza in Chapter 8 with her months and months of useless treatment? The problem is, it's not so easy to put this into practice when well-equipped diagnostic laboratories aren't always available. Even when they are, it can take months for the results to come back. Case in point: US lawyer Andrew Speaker, who flew across the world with suspected XDR-TB. His diagnostic timeline started out in January 2010 with a strange lung mass on an X-ray taken for an unrelated problem. Over the next few months, he had a couple of sputum smear tests but they came back negative. As we know, though, sputum smear microscopy doesn't catch every case. Especially not people like Speaker, who wasn't outwardly sick. So a positive TB diagnosis couldn't be made until his sputum samples grew up in a culture tube. It was not until May – four months after that initial X-ray – that a diagnosis of MDR-TB was finally proposed. At the end of May, after Speaker had already flown to Europe, the CDC lab finally decided that his strain was actually XDR-TB. Everyone freaks out blah, blah, blah. If it takes the US months to nail down a drug-susceptibility test, imagine how complicated the process is in TB-endemic countries without the CDC's level of funding?

The issue of diagnosing TB in low-resource countries is an ongoing problem. But over the last decade or so, the diagnostics landscape has been upended by an unassuming piece of tech that wouldn't look out of place in the kitchen. The GeneXpert MTB/RIF looks like an espresso machine designed by someone with no flair for design. Boxy. Grey. With little flaps on the front into which small blue cassettes are placed; a bit like those ground-coffee capsules, only they're slightly bigger and not advertised by George Clooney. And in the case of GeneXpert, the cassettes contain human sputum. GeneXpert MTB/RIF is a self-contained unit that detects *M. tuberculosis* DNA. The machine uses PCR to amplify a sequence of the *rpoB* gene, which is specific to *M. tuberculosis*. That's the diagnosis part. It also uses molecular beacons to look for mutations in *rpoB* indicative of rifampicin resistance. This is where it's a really clever piece of kit. Two hours, and not only do you have a yes or no diagnosis, but you also have a pretty good idea if the strain responsible is a multidrug-resistant one. If it's positive, the sputum sample can then be sent for more in-depth drug-susceptibility testing using the older and slower but more comprehensive methods.

When the GeneXpert MTB/RIF appeared during the early noughties, phrases like 'magic bullet' and 'revolutionary' started to be tossed around. It is, after all, the biggest breakthrough development in TB diagnostics since Robert Koch's unfortunate discovery of tuberculin. In a 2010 paper published in the *New England Journal of Medicine*, the authors described how their assay could identify 92 per cent of patients (compared to the gold standard of culturing bacteria from sputum). This included 73 per cent of patients whose infections had been missed by sputum smear microscopy but later came up positive by culture (these TB cases are referred to as sputum-smear-negative). When it came to detecting rifampicin resistance, the assay was correct around 98 per cent of the time, again using culture as a comparison. So, not perfect, but a heck of a lot quicker with the two-hour turnaround time, and a vast improvement on sputum smear microscopy. To drum home the numbers, of 171 patients who were diagnosed by microscope as not having TB, 124 were positive when tested once with GeneXpert MTB/RIF, rising to 154 after three tests. That's 154 people who could be started on TB therapy straight away rather than having to wait for the results of the culture-based test. The GeneXpert assay wasn't the first PCR-based test for *M. tuberculosis* but it was the first one that got everyone ridiculously excited.

Without FIND's work in developing and implementing GeneXpert for TB, the technology would probably be confined to US post offices looking for anthrax in the mail. We need organisations like FIND that can navigate the worlds of industry, academia, politics and people to make big things happen. Claudia describes FIND as a catalyst for building partnerships. A TB diagnostics matchmaker, I suppose, able to bring all the right people and the right information together to benefit those most in need. When it came to GeneXpert, FIND were navigating unexplored ground, so it's perhaps not too surprising that things didn't go exactly to plan. 'It has not had the scale-up and the impact that we had hoped it would have,' Claudia told me, and a part of this is down to the price. In 2010, the WHO endorsed GeneXpert MTB/RIF, and so began a worldwide rollout of the technology, with concessional pricing for low- and middle-income countries. In 2014, 18,000 test modules and 10 million cartridges were procured. But despite the price reductions, the technology is still far more expensive than sputum smear microscopy. A 2011 paper looked at the comparable costs of different diagnostics in India, South Africa and Uganda at the start of the rollout. Overall, each test (including all costs such as cartridges, equipment and salaries) ranged from $14.93 to $27.55 depending on the setting and how frequently the test could be used. This was in comparison to $1.13 to $1.63 for sputum smear microscopy. The budget simply isn't there in most countries, no matter how cost-effective the test. There are also questions over what gap in the TB diagnostics landscape GeneXpert is trying to fill.

Today, GeneXpert is recommended by the WHO as the initial diagnostic test in all individuals presumed to have pulmonary TB. Resources permitting, it may be used to replace microscopy altogether as the initial test. Catherine Cosgrove and Onn Min Kon both told me how, at their London hospitals, GeneXpert is used immediately on any sputum sample that is clearly positive for TB by sputum smear microscopy. In other patients, the test is performed should a culture grow up from their sputum, giving a quick indication of whether it is a drug-resistant infection while the team wait for the subsequent culture-based drug-susceptibility testing. A friend of mine who works in another London diagnostics laboratory told me that they rarely bother with microscopy, instead sticking most samples straight into the GeneXpert (their budget must be higher than St George's and St Mary's). Everyone still relies on culture for a definitive answer, but GeneXpert means doctors can be more confident that they've started the right patients on the right therapy. But – and this is a big but – sick patients are already being started on treatment prior to diagnostic confirmation. Yes, doctors get to have a slightly difficult conversation with a patient should the test come back negative, but it's a better option than letting someone die while you wait for results. This makes me wonder how much impact GeneXpert can have on improving diagnosis rates, and not just in high-income countries such as England but also high-TB-burden settings.

In 2016, a group led by David Dowdy of the John Hopkins Bloomberg School of Public Health independently evaluated the scale-up of GeneXpert when used in real-life situations across Uganda. Many previous studies had looked at its efficacy in clinical trials and in middle-income countries such as South Africa. Uganda, in comparison, is right up there with the poorest places in the world and is among the WHO's 22 high-TB-burden countries. In 2012, its diagnostic approach was sputum smear microscopy followed by a chest X-ray in symptomatic patients who came up negative. Often, the limitations in diagnosis meant that patients had long delays in accessing treatment, and this was hitting HIV-infected individuals particularly hard. So in 2012 international funding was procured to set up GeneXpert in a number of facilities to enable diagnosis of HIV-positive TB patients. As of June 2015, a total of 10,013 HIV-positive patients had been tested with GeneXpert and 1,494 diagnosed with TB. The recommendations for GeneXpert use are in smear-negative-HIV patients, children, healthcare workers or those who had been in contact with MDR-TB cases. David Dowdy's study, however, suggested that these instructions were not quite filtering down to the people performing the tests. The team reviewed records from 18 healthcare facilities, some with GeneXpert capacity and some without. One of the authors' conclusions was that GeneXpert is greatly under-utilised in Uganda, with only 8 per cent of the total testing capacity being put to use, equating to around one GeneXpert assay per work day in each facility. Only 21 per cent of sputum-smear-negative patients with suspected TB received the GeneXpert test – it should have been all of them. Running out of cartridges, module malfunction and high error rates were commonly reported problems.

Other implementation studies have found similar problems. In Mozambique and Swaziland, GeneXpert implementation came as a package deal with dedicated specimen-transport networks, supply networks, laboratory training and process support. So, all the infrastructure needed to give those poor neglected GeneXpert machines the best chance of actually being used. Despite all this, both countries were still only using two-thirds of their GeneXpert capacity by 2015. Even with the underuse of the test, Mozambique did achieve a 69 per cent increase in cases of TB detected. However, it had a huge amount of outside support. The results from David's Uganda study show that implementing a brand new technology without the necessary infrastructure, additional staff and process support is unlikely to improve patient outcomes. There's little point placing a GeneXpert machine in a lab where the underpaid and overworked staff have received the minimum of training, and where only one or two tests will be performed per day. Sure, the assay may take only two hours, but who is then going to ensure that those results are sent back to the right patient? Or track that patient down if they have hopped on the bus back home? Or work out how to fix the machine when someone rams a cartridge in upside-down? Basically, in some settings, the machines are just sitting there like big ugly espresso machines that no one really knows how to use. Even if someone does get the urge to brew some coffee, George Clooney has used the last cassette and not put in a new order. Next year will see the release of the GeneXpert Omni – a new and improved version that addresses some of the implementation problems observed with the original. 'GeneXpert, at the time, was designed the best that it could have been,' Claudia told me. 'We're taking the learning to make the instrument more rugged. You can blow dust at it all day long, as happens in countries where TB is highly prevalent … you can be at 40 degrees centigrade [104 Fahrenheit], and the reagents are still stable … you can drop it … and it can be very, very simple to operate.' Big improvements, but will they make a big difference? Not unless healthcare system improvements come along with it. Things like ensuring patients get tests, machines are maintained and healthcare providers are using the results correctly.

I talked to David Dowdy about why GeneXpert hasn't been the huge transformative change that many hoped it would be. In his study, 19 per cent of the tests that were performed came back positive, so it's clearly successful at diagnosing TB. However, the use of the machine didn't seem to lead to any increase in patients being started on TB drugs or a decrease in the time taken to begin treatment. David explained how doctors in settings like Uganda, just as in England, were already diagnosing TB on their own without relying on sputum smear microscopy tests. Few wait for culture results. If a patient is sick with what looks like TB, they're put on treatment. So for the majority of patients, GeneXpert – as it stands – isn't going to change how their TB is managed. I don't want to suggest that it hasn't been a success on many levels, though. India, for example, is currently scaling up its GeneXpert capacity, partially as a response to its previously undocumented levels of drug resistance. For individual patients like South African Phumeza Tisile with no obvious MDR-TB risk factors, GeneXpert can make a big difference in reducing the delay before starting individualised treatment. I don't doubt that India's hopes for bringing drug-resistant TB under control will rely on detecting cases a lot faster than using old-fashioned methods. And one of the huge advantages of GeneXpert is that the relatively quick turnaround compared to culture means that there's less time for a patient to get lost after visiting the clinic. For these people – most likely the ones

who aren't seriously ill (yet), with a negative sputum smear –
GeneXpert fills the same gap as those giant rats, only with the
added extra that it can pick up on drug resistance. 'If I were a
patient I would want GeneXpert, not just a smear,' David told
me. 'It's a test that's good for individual patients and we should
be using it, but it's not necessarily something that's
revolutionised the public-health approach to TB.'

GeneXpert is not (and doesn't claim to be) a replacement
for culture-based drug-sensitivity testing. For starters, it only
detects resistance to one antibiotic – rifampicin. There are
other molecular diagnostics out there, including the
MTBDR*sl* line probe assay, which can detect a number of
mutations. But MTBDR*sl* doesn't do a brilliant job at
identifying resistance to certain drugs, highlighting one big
problem with molecular drug-susceptibility testing. It's
impossible to test for the presence of drug-resistance mutations
when we do not know all of the mutations responsible for
resistance. Rifampicin resistance is predominantly the result
of mutations within a small region of *rpoB*, with 95 per cent
of mutations explainable by changes to this section of DNA.
Isoniazid is slightly more complicated, with only around 64
per cent of observed resistance mutations occurring at a
specific position in the *katG* gene. The second most common
mutation is in the *inhA* gene and occurs in 19 per cent of
isolated strains. Combined with the other known mutations
conferring resistance to isoniazid, only 84 per cent of the
global resistant strains can be explained. On top of that,
different geographic distributions of resistance mutations
mean tests may work in one part of the world and not in
another. In 2015, it was reported that rifampicin-resistant
strains in Swaziland had a surprisingly high rate of a usually
rare mutation. The authors of the study looked at strains
isolated during a national survey of drug resistance and found
that 38 of 125 MDR strains would not be detected by
GeneXpert MTB/RIF, making the test very unreliable in
this one setting. I'd personally like to see a new version of
GeneXpert that looks for additional resistance mutations in
rifampicin and isoniazid, allowing it to flag up a larger

proportion of MDR-TB cases. Could we develop GeneXpert-like tests that go further to detect additional mutations in any number of genes? With a greater knowledge of resistance-causing mutations, yes. But the closer you bring a test to the patient, the less use a more in-depth drug-susceptibility testing actually is. There's not much need for a full drug-susceptibility profile at the point of care, when treatment of these more complex cases needs someone with more expertise to design drug regimens. 'However, if we have two or maybe three regimens that come in blister packs and are easy to take in five years' time, then we need more DST at lower levels as well,' Claudia Denkinger told me. 'And given that it takes five to ten years minimum to bring a test to the market, we need to think about it now.'

Molecular tests undoubtedly have a huge role in the future of drug-susceptibility testing. One thing that many of the people I spoke to agreed on was that, one day, whole genome sequencing will be a key part of TB drug resistance testing (when is another matter). It's a technique that yields vast amounts of information. Useless information if no one knows how to interpret it. Is that mutation conferring drug resistance or is it a random mistake that has no impact on the bacterium? And do we need to worry about low-level drug resistance? Without a full dictionary of what every resistance mutation looks like, it's difficult to make sense of all those long strings of Gs, Ts, As and Cs. Whole genome sequencing, however, may be its own saviour here, by providing the information required to better understand how resistance emerges.

One of the first papers to look at the potential of whole genome sequencing for drug-resistance testing of *M. tuberculosis* was published in 2014. The authors describe the treatment of a 38-year-old man first admitted to a UK hospital with clinical and X-ray symptoms of TB. His sputum smear results were negative but a bronchoalveolar lavage (basically, the lungs are rinsed out with saline to look for bacteria) came back positive. So the doctors started him on standard first-line treatment. All good so far, and the patient was discharged after two weeks. Only, then the cultured bacteria from his lungs started to grow and, as is usual, the reference laboratory worked out its genotype and drug-resistant profile. It was a Beijing strain resistant to rifampicin, isoniazid, ethambutol, prothionamide, pyrazinamide, streptomycin, moxifloxacin and ofloxacin. Oh dear. And to make things worse, the patient was of no fixed abode and had vanished. Finally, three months after he'd first been admitted, he was detained under Part 2A of the UK Public Health Act 1984 and admitted into an isolation unit at a Cambridge hospital. His treatment was hampered by side effects, poor compliance and – most importantly – the long time taken for those drug-susceptibility tests to come back. He was finally discharged after 184 days.

But if whole genome sequencing had been on the scene? Well, the scientists led by Professor Sharon Peacock of Cambridge University went back to that first cultured strain and sequenced it. They looked at known genes associated with resistance to 39 drugs. Nine matched up with the culture results from the reference laboratory. They also found mutations consistent with resistance to a number of other drugs that hadn't been tested in real life. Had this technique been applied to the patient's initial strain, the time taken to identify his infection and start him on the correct drugs would have been dramatically shortened. But of course it's easy to say that in retrospect. The problem is knowing what patients to send for this expensive testing when the majority don't have XDR-TB. A 2016 study looked at the rationing of drug-susceptibility testing in Lima based on the WHO's recommendations for who should receive GeneXpert testing. More than half of those with drug-resistant TB had no risk factors and would therefore have not been tested.

I personally agree with David Dowdy's belief that drug-susceptibility testing shouldn't be rationed. 'I don't think we have to test everyone who has symptoms of TB for drug-susceptibility, because 90 per cent of them won't have TB,' he said. 'My view would be you can start anyone on treatment while you wait for drug-susceptibility test results. But you should have that drug-susceptibility test for everyone who has a confirmed diagnosis where you have the capacity for it.'

Speeding up the test is going to be a big part of making sure the results get back to the patient in time to help. One day, I imagine that whole genome sequencing of an infecting strain will be performed directly from a patient's sputum. The doctor will be able to view the results with user-friendly software that not only identifies all the major resistance mutations but can predict how best to interpret this information into a successful drug regimen. Of course, a sticking point here is the fact that we only have one standard cocktail of drugs available right now, and treatment of drug-resistant cases involves more of an individualised, whatever-is-available approach that requires an expert to design. I'll come back to drugs, or lack of, in the next chapter.

What I've focused on for all of this chapter is the science and the technology side of things. However, this is only part of the solution. Three million TB cases go undiagnosed a year, but these aren't all in patients who are coming to the clinic and being missed by sputum smear microscopy or Mandy the rat, or even an out-of-order GeneXpert machine. Many of these patients aren't getting to the clinic in the first place; maybe because they have just a mild cough that doesn't seem too serious, or they live miles from the hospital so they talk to their chemist instead, or maybe because their religious beliefs send them to a traditional healer rather than a medical doctor. David explained that we need ways of capturing patients at an early stage before they get ill and transmit TB to others. He thinks that we should be 'raising awareness of the importance of testing in general. Like we have people being tested for HIV not because they have symptoms, but just because they might have HIV. So increasing the awareness of the need to test and going out into the community to find people who have TB and, among those who've been infected, giving them preventative therapy to prevent progression.'

The answer to the problem isn't going to come from new technologies alone. In fact, part of the solution could be as simple as knocking on someone's door and asking 'Have you been coughing for more than two weeks?' It's part of an active case-finding approach that's recently been trialed in India

– home to a third of those 'missing 3 million'. Some of these lost patients are hidden among the vast urban slums; others are isolated by geographical or cultural factors. A 2016 paper summarized efforts to find TB across 5 million Indian households using trained community volunteers. In all, over 350,000 potential cases were picked up from more than 20 million people. Getting these people tested for TB, though? That was where things got messier. Of the potential cases, only 22 per cent ended up at the microscopy centre. Solutions include physically accompanying patients to the clinic or collecting and transporting their sputum samples minus the patient. But what is very obvious is that relying on passive case-finding in which TB sufferers arrive at the healthcare centre under their own initiative does not work. We need new and inventive ways of ensuring that no one slips through the net, and that everyone who needs a diagnosis receives one. 'One of the things that we're looking into is incentive-based approaches to case-finding,' David told me. 'So having people with TB or at high risk of TB going out into the community and recruiting their friends or colleagues or people who they think have TB to come in and get tested for a small incentive. Do I think this will be the magic bullet that will transform TB control? No. But I do feel that it has the potential to bring in and detect people with earlier forms of TB that would otherwise not be diagnosed through the healthcare system.' A theme of my conversation with David was that, barring an unforeseen breakthrough, TB diagnostics are going to continue to advance incrementally. There's no one answer to the problem, and what we need is a comprehensive approach. More and more, people are coming to recognise that focusing on the social barriers to diagnosis is integral to finding those missing 3 million. It's easy to say that, though. Putting it into practice is something else.

CHAPTER THIRTEEN

New Drugs for Bad Bugs

'I admit there was a bit of luck involved,' Koen Andries tells me. He's talking about the development of bedaquiline, which in 2012 became the first new TB drug approved in more than 40 years. Few scientists can honestly say that their work is going to save thousands – maybe hundreds of thousands – of lives. So part of me was expecting Koen to be the scientific equivalent of a rock star. Own line of pH buffer, designer lab coat, that sort of thing. Instead, he is the epitome of restrained calm and professionalism, quick to give credit to the other members of his team and lacking in any shred of arrogance. When I asked him how many people had been treated with bedaquiline, he told me it was about 10,000, most of them pre-XDR and XDR-TB patients. Ten thousand people! Many of whom may have died if it hadn't been for Koen and his team's work. What makes it an even more remarkable story is that without his unwavering – perhaps even stubborn – belief in the drug (plus that big dose of good luck), bedaquiline wouldn't have made it. 'When I tell the story about the development of bedaquiline,' he says, 'I call it a rocky road. There were really *big* rocks on the road.'

Back in the late 1990s, a colleague of Koen's at Janssen Pharmaceutica happened to hear about the WHO's then-announcement that TB had once again risen to become a global health crisis. He also happened to have some spare wells in the microtiter plates his team were using to look for new compounds capable of killing fungi and other undesirables. He didn't have access to the containment laboratories required to work with *M. tuberculosis*, so he picked a non-pathogenic species of mycobacteria – *M. smegmatis* – and had his people include it in their screens. Every time they ran a plate of potential new drugs, they tested them against *M. smegmatis* in the otherwise empty wells. At the time, screening compounds against *M. smegmatis* was a laughable approach in the TB drug-discovery world. Everyone was all about new and shiny target-based drug discovery, not old-fashioned whole-cell-based screening. And if you were going to go all retro and use whole cells, then at least use *M. tuberculosis*, for Koch's sake. Because what's the point in finding something that can kill *M. smegmatis* – a non-pathogenic environmental species that, at a push, might infect the odd syphilitic penis or anabolic steroid injection site? All that work, and your newly discovered compounds may not even work against *M. tuberculosis*. Luckily for the bedaquiline story, the overall opinion of the majority of the TB drug-discovery field didn't matter, as the Janssen *M. smegmatis* screen wasn't official. It was more of a pet project performed without the need for extra funding or approval and, had it been unsuccessful, the rest of the world would never have known it had taken place. If the scientists involved had been required to gather external expert opinion on the project, it would never have happened.

So the pharmaceutical wheels continued to turn and everyone went on with their jobs and lives. The tally of compounds screened rose and rose, until more than 10,000 had been tested. Finally, a year or two after the first wells were inoculated with *M. smegmatis*, something happened. Or, rather, it didn't happen. The wells containing *M. smegmatis* didn't grow. The team had discovered a starting point for what would eventually become bedaquiline. At the time, everyone was more interested in the fact that the new compounds could also kill *H. pylori*. This species had only just been discovered to be the main cause of stomach ulcers – a big market for new drugs. TB, though? Well, let's just say that TB drugs aren't particularly profitable. And anyway, efforts to confirm that the compounds worked against *M. tuberculosis* had yielded some conflicting data. Plus they weren't very *nice* compounds. Insoluble, quite toxic, didn't look much like a drug. This all added up to many people at the company becoming highly uninterested in continuing to develop bedaquiline. Koen Andries, though, had seen something special in the compound. He wasn't a trained bacteriologist, but his background in veterinary medicine meant he recognised just how big an impact it could have against TB. So he kept pushing.

The team found the mechanism of action – a unique inhibition of the cell's energy-generation pathways. They worked on new formulations and tested them on mice. Against all the odds, one of the new versions worked better than the existing TB drugs. It was almost unbelievable, to the point that, when the team published a *Science* paper detailing what was then known as TMC207's spectacular potential as a new TB drug, many of the haters continued to hate. But while some scientists were sceptical, others were starting to get very excited. I was working in the TB drug-discovery field at the time. I remember thinking how the *Science* paper made it all sound so easy, and I couldn't help but feel a little jealous of all the good luck that seemed to have blessed this project. It's only now that I realise luck means nothing if someone's not prepared to grab hold of it and refuse to let go. 'One of the take-home lessons is that you need people like myself,' Andries told me. 'You need a few people who are behind the drug, who defend the drug. Because otherwise, with the slightest setback, the development of the drug will be stopped. It was very close in this case, on several occasions. Bedaquiline barely made it.'

One of the biggest rocks in the road came after the *Science* paper, when the team embarked on the first clinical studies in humans. The way these studies work is that TB patients are given the drug by itself for (usually) just two days and you look to see if the numbers of bacteria in the lungs go down. Bedaquiline didn't work very well. Even after one week, the compound had only just managed to kill some of the bacteria, and at a level far below that seen in the control patients given either rifampicin or isoniazid. In humans, unlike in mice, the old drugs were much better than the new drug, and no one knew why. The drug also turned out to remain in the human body for a very long time, something that made some people very nervous. But Andries persisted in pushing the drug forward, shoving it over a couple more bumps in the road

until some new human trials yielded the kind of data that
can't be argued with. Bedaquiline might work slower than
other drugs at first, but it does work if you're patient. And
when it does, the results are dramatic. Finally, more than 15
years after the project's conception, bedaquiline was granted
FDA approval. It's an approval with caveats, what with one of
the side effects of the drug potentially being death. In trials,
while more people were cured in the bedaquiline group
compared to the placebo, more of this group also died. It's
worth noting that this could well have been a fluke resulting
from the relatively small numbers of patients included in the
trial. But until bedaquiline proves itself, it is only to be used
when a complete regimen cannot be cobbled together from
other drugs and patients really have reached the point where
'kill or cure' is an option. Setting it aside for drug-resistant
cases for now, though, is actually a good thing. Because if
we're not careful in how bedaquiline is unleashed upon the
world, we could easily end up selecting for resistant strains of
M. tuberculosis. 'It's not easy to discover and develop a drug
such as bedaquiline,' Andries said, talking about how the
drug should be used in the field. 'It only happens once in 20
years, and if you don't do it right, you can lose the drug in
much less than 20 years.'

Twenty years is a long time, but when bedaquiline was first
unleashed upon the scientific community, that part of the
message got a bit lost in all the excitement. TMC207 arrived
at a time when everyone's patience in target-based drug
discovery was beginning to wear thin. The genomic
revolution was supposed to make everything easy. Pick an
important target, purify the protein, develop an assay, screen
for compounds that can stop the protein from working. Only
we all tried and tried, me included, and got nowhere. And
then bedaquiline came along and seemingly proved that the
old-fashioned whole-cell screening methods were the best
way forward after all. The project I was working on was
already very sick by this point, but the field's pendulum swing
from target-based back to phenotypic whole-cell screens was
among the final blows that killed it. On many levels, the

proponents of testing compounds against actual bacteria – be it *M. smegmatis* or *M. tuberculosis* – have a really good point. Koen Andries told me, 'When you do phenotypic screening, you're testing 614 targets at the same time. Because there are 614 essential targets in TB. So you put one drug in a test plate and look whether the TB bacilli grow or not – if it doesn't grow then [the drug] affects one of the 614 targets. If you do target-based research you're looking at just one target.' What the bedaquiline discovery pathway did well was to cherry-pick the best parts of whole-cell and target-based screening. I would say many scientists would agree that this is where the future of TB drug discovery lies. One such integrated method is to start with the whole cells but look for the target immediately after finding those first hits, then continue with a target-based approach. An alternative is to introduce a target-based element into the whole-cell screen, for example finding a readout of your target protein's activity that can be measured from outside the cell. Clever approaches, but that doesn't mean they work every time.

Following bedaquiline's discovery, lots of TB researchers returned to phenotypic, whole-cell screening in the hope that they could repeat this road to drug-discovery greatness. Bedaquiline hasn't been an easy find, though, and many, many whole-cell approaches – like their target-based counterparts – end up on the scientific scrapheaps of unpublished, unsuccessful projects. Koen Andries explained how the success of a screening programme – whether whole-cell or target-based – is built on the strength of the initial compounds. 'What most people under-appreciate is that in order to find the needle in the haystack, you need to find a good haystack first. Not every haystack contains a needle! And some haystacks may contain more than one.' He had access to what he describes as one of the best haystacks in the world, made under the direction of Paul Janssen – 'the best medicinal chemist ever'. The quality of that haystack is all the more important in TB drug discovery thanks to that big fat waxy wall making it difficult to get compounds inside the cell. Before you've even started, your hit rate is going to be lower than for other disease areas such as malaria.

Even if you do obtain promising hits, there are plenty of other rocks in the road. Back in 2010, a paper from Thomas Dick's lab at the National University of Singapore caught my eye by virtue of having the words 'devoid of *in vivo* efficacy' in its title. In lay-person speak, this translates as 'doesn't work in the host'. This was a *Nature Communications* paper, yet based on the title, it was describing work that had failed quite spectacularly in its initial aim. Negative results aren't often published (but should be). The project had started with a whole-cell screen. A few lead compounds were identified as capable of preventing bacterial growth and a lead optimisation programme used to generate 324 derivatives. The team tested some of the compounds in a mouse model and ... nothing happened. In animals treated with isoniazid, the numbers of bacteria in the lungs had dropped dramatically by 20 days. The new compounds, in comparison, might as well have not been there. The problem, it turned out, was down to how the bacteria were grown for the screen. Remember how I talked about how we use glycerol as a carbon source in the lab, whereas *M. tuberculosis* actually likes to feast upon lipids in the host? Well, it turns out that the compounds isolated during this project were interfering with glycerol metabolism, causing the cells to poison themselves. Remove glycerol, though, and the compounds do absolutely nothing. I mention this study because it underscores how we need to understand how *M. tuberculosis* survives in the host in order to be able to kill it.

A really interesting area of TB drug discovery tries to take what happens inside the host into consideration by testing how well a new drug can penetrate a TB lesion. It's not the easiest structure to gain access, and poor penetration could lead to sub-killing concentrations of a drug reaching the bacteria. So one team of scientists recruited 15 patients all scheduled to have parts of their lungs removed, gave them some TB drugs prior to surgery, then afterwards used matrix-assisted laser desorption/ionization (MALDI) mass spectrometry to create a 2D image of each lesion showing the concentration of each drug. A TB lesion isn't a uniform environment. It's like a little town, inhabited by lots of bacteria potentially doing slightly different things. You've got your bacteria surviving inside macrophages of various activation states. You've got bacteria pushed into a state of slow or non-growth by areas of hypoxia. You've got bacteria surviving outside of the host cells within the cheesy core of a cavitating granuloma. Growing bacteria, dividing bacteria, sleeping bacteria, barely alive bacteria. Any cocktail of TB drugs needs to be able to not only kill the different populations, but to get to them in the first place. The 2015 *Nature Medicine* paper from Véronique Dartois's lab at Rutgers University showed how different drugs behave very differently within the same lesion. Pyrazinamide, for example, was seen to be distributed through the lesions, finding its way into the cheesy caseum as well as the parts rich in infected macrophages. Moxifloxacin, however, didn't reach high enough concentrations in the caseum, perhaps explaining why this drug hasn't performed brilliantly in clinical trials. Dartois's lab has since designed a way to predict how well a drug will penetrate a TB lesion. In the future, testing if drugs are able to diffuse all the way to the centre of a granuloma will hopefully happen at an early stage of drug development. No more losing potential drugs because they behave differently in a host than in the lab.

There's lots of exciting work happening at the early stages of drug discovery. There's even a really interesting partnership effort under way called the TB Drug Accelerator Program, which is a collaboration between various pharma companies and research institutes, with funding from the Bill & Melinda Gates Foundation. The idea is to help early-stage drug discovery along by getting everyone to talk to each other and share compounds and findings, all in the hope that they'll be able to come up with a TB drug regimen that can cure patients in just one month. I'm actually quite optimistic about the early discovery efforts under way in multiple labs around the world. There's certainly a lot of really interesting and inventive science going on; I'm just not so sure about the rate at which this is being converted into actual drugs capable of saving lives. In 2012, bedaquiline wasn't the only drug to be granted conditional approval for use in MDR-TB patients. It was joined by another brand new drug, delamanid. Unfortunately, two new drugs in such a short space of time are not indicative of a full-to-bursting TB drug-development pipeline, poised to gush out lots of new alternatives to the existing, resistance-ridden drugs. In fact, the TB drug-discovery landscape is more the Algerian desert than the Amazon. Dry, parched, somewhere you probably don't want to picnic.

Because there are so few options currently in human trials, it's a big setback when trials are put on hold due to side effects, or candidates are pulled because they're not working very well, or bureaucratic delays slow everything down. I can cover everything new in just one paragraph, in fact. First, there's a drug called pretomanid currently in phase III trials. Pretomanid was discovered through a project to scour the literature for anti-TB compounds that had been abandoned due to some nasty properties, and then make them better. This is a very different approach from the expensive screening of tens of thousands of compounds that's been used in other drug-discovery attempts, and it represents a really clever way of finding a new TB drug on a budget. However, at the time of writing, the STAND (Shortening Treatments by Advancing Novel Drugs) trial looking at pretomanid in combination with moxifloxacin and pyrazinamide had stalled due to high levels of liver toxicity among the participants. It's due to start up again soon, though, with some additional safety procedures in place. A new player, Q203, entered clinical testing in 2015, but the results won't be out until 2017; a more advanced candidate, sutezolid, seems to be stuck somewhere in phase IIa trials thanks to a combination of intellectual property negotiations and other non-science hold-ups. And that's it for the completely new drugs in the later stages of clinical development. Bedaquiline, delamanid, pretomanid, sutezolid and Q203 (which doesn't even have an annoyingly hard to pronounce name of its own yet). There are others teetering on the edge of clinical development (look out for the benzothiazinones in the future), but I'm not going into detail here.

It's not all about the new drugs on the block, though. There's also work under way to optimise the use of existing drug families or repurpose drugs already licensed for other therapeutic areas. Linezolid, for example, is already used off-label for MDR-TB (such as Phumeza Tisile's infection), as is clofazimine – a drug licensed for leprosy treatment. These drugs, as well as bedaquiline and delamanid, where clinical trials have not yet been completed, are what's considered 'group 5' drugs by the WHO, due to a lack of sufficient data on their efficacy and toxicity. They can be used if there are no other options – when it comes to treating people who will otherwise die, the rules and regulations are less restrictive. One example of this repurposing approach is thioridazine – a drug licensed as an antipsychotic medicine, not an antibiotic. I'm including its story here not because it is going to solve the TB problem but because it's a case study of how scientific discoveries need someone to spot their significance and chivvy them towards the finishing line. Years ago, I spent several months screening a whole library of known drugs against *M. tuberculosis*. Drugs already approved for use in other conditions come with the advantage that a lot of the testing has already been done – a bit like a boy band member forging a solo career. I did it blind in that I had no idea what was in each test well until after the experiments. And I found lots of compounds that could kill TB! But most of them proved to be actual TB drugs, which was both disappointing and reassuring. Among the others were a malaria drug, a dye called methylene blue and, surprisingly, an antipsychotic medication used to treat schizophrenia and other psychiatric disorders. This drug didn't kill *M. tuberculosis* that well, though, and had some issues with reproducibility, and I wasn't about to try to use my chemical ordering privileges to buy a highly controlled drug that the safety department would have had a field day with. In any case, we had other leads to follow so nothing more came of out discovery. But in writing this book I've come to realise that the potential of thioridazine and its derivatives had actually been noticed as far back as the 1950s.

Back in the days of Robert Koch, scientists were playing around with ways to stain bacteria so that they were easy to visualise in infected tissue. One of the dyes used was called methylene blue, a member of the phenothiazine family of compounds. Then in 1899, an Italian doctor called Pietro Bodoni used methylene blue on his psychiatric patients and suggested that maybe it could have a use in the treatment of numerous psychotic conditions. Despite its potential, though, it never made it into widespread use, partly because it couldn't be patented.

As the nineteenth century rolled into the early twentieth, other uses for phenothiazines were discovered. An antihistamine for the treatment of motion sickness, a drug to combat circulatory shock during surgery, a potentiator for general anaesthetics. Then, in 1952, came clinical trials of a phenothiazine called chlorpromazine as a miracle drug for psychiatry. It succeeded where methylene blue had floundered. What with it being the 1950s, some of those treated with chlorpromazine also had TB, and interest in the bacteria-killing properties of the phenothiazines was kindled. The problem in their use to treat bacterial infections was that they have some severe side effects if used for a prolonged period, such as involuntary movements of the face and jaw. So when isoniazid appeared on the market, further development of phenothiazines as potential TB drugs slowed. But throughout the years, various scientists have repeatedly championed the idea that these compounds can kill *M. tuberculosis* – all the way from research published in 1956 indicating that TB patients being treated for psychiatric disorders showed an improvement in their infections, through to today when scientists remain interested in the biotic aspect of these 'narcobiotics'. Today, research focuses on thioridazine, as it can be used to kill *M. tuberculosis* at a concentration low enough to prevent the side effects. It has, in fact, been used on a compassionate basis to treat patients with XDR-TB and no other options.

Zarir Udwadia, the doctor who highlighted India's drug-resistance problem in dramatic fashion, was part of a small study looking at the use of thioridazine as part of what's known as salvage therapy for XDR-TB. Four patients were given thioridazine on top of the four-drug regimen decided upon from whatever TB drugs were still available to them. All of them took linezolid, clofazimine and cycloserine with thioridazine, plus clarithromycin, capreomycin or PAS, depending on their resistance profile. The paper reports that three of the four patients showed some clinical improvement and the drug was well tolerated in terms of side effects. It wasn't a happy ending for any of the patients, though. One died after three months, another after five and a third after seven. The fourth patient was 'lost to follow-up' but was showing no signs of improvement after four months of therapy, so it's unlikely he recovered.

Argentinian attempts to use the drug were more successful. Thirteen XDR-TB patients were treated with linezolid, moxifloxacin and thioridazine in addition to a couple of other agents where possible. Many of the patients were cured, although it's not possible to say how much thioridazine contributed to this. There are no negative controls here – withholding a potentially life-saving drug from some patients is obviously unethical. But studies like this, as well as some impressive work in the laboratory showing that thioridazine is effective against *M. tuberculosis* and can promote macrophage-mediated killing of the bacteria, raise the question of whether thioridazine should be tested in controlled trials. Leonard Amaral is possibly the drug's biggest supporter, having published extensively on its use as a TB drug. I asked him what the barriers are to the use of thioridazine as a TB therapy. 'Money cannot be made given that TZ [Thiorizadine] is no longer under patent protection,' he explained, reminding me of the methylene blue story. He believes that while thioridazine 'is cheap and safe to use when used by a competent physician', it will never become a permanent part of TB control strategies without the 'political will to cure and control TB infections'.

Wherever they come from, we clearly need more TB drugs to solve the problem of resistance. We'd also like to improve upon those long, long treatment times using complicated regimens fraught with unpleasant side effects. The ideal regimen? One that is simple, with the minimum number of agents. One that is fast. An all-oral regimen that avoids side-effect-prone injectables and takes some of the pressure off the healthcare system. Readily available, cheap drugs. A single cocktail that can be used for not just MDR-TB but also drug-susceptible TB. So let's look at the trials currently under way aiming to get us a bit closer to this pipe dream. There's some interesting work focusing on using high doses of the rifamycins, of which cornerstone TB drug rifampicin is a member. There's one trial – TRUNCATE-TB (Two-month Regimens Using Novel Combinations to Augment Treatment Effectiveness for drug-sensitive Tuberculosis) – due to begin fairly soon, looking at speeding up treatment to just two months using high-dose rifampin in combination with new and existing drugs. TRUNCATE-TB is among just a few trials looking at shortening treatment duration for drug-sensitive TB; much of the work focuses on reducing the two-year treatment time frames involved in curing drug-resistant TB. It's more of a pressing problem than drug-sensitive TB, what with only 10 per cent of worldwide drug-resistant TB cases currently being treated adequately and the survival rate for XDR-TB as low as 15 per cent.

This year, we saw the WHO approve what's known as the Bangladesh regimen for MDR-TB. The good news is that it works in just 9 to 12 months and costs a third of the price tag of the original. Early data from the STREAM (Standardised Treatment REgimen of Anti-tuberculosis drugs for patients with Multidrug-resistant tuberculosis) trial, including a higher than 80 per cent cure rate, were promising enough for the WHO to recommend the new regimen before the official trial results were even available. The bad news? It involves seven drugs – moxifloxacin, clofazimine, pyrazinamide and ethambutol taken for nine months with the addition of kanamycin, high-dose isoniazid and prothionamide for the first four months. It's also not suitable for anyone with resistance to second-line injectables (kanamycin, for example) or a fluoroquinolone, so pre-XDR and XDR-TB are ruled out. A continuation of the STREAM trial – STREAM2 – is currently enrolling patients to look at adding bedaquiline to this regimen and includes a study arm that replaces kanamycin to form an all-oral regimen. But is this approach of adding new drugs to lots of other toxic drugs the revolution in TB treatment that we need? Or is it a big improvement but still an untenable solution in the long run? In some settings, early discontinuation of existing treatment regimens is as high as 50 per cent. These new combinations seem a lot like more of the same to me, although should STREAM2 be successful, the brutal drug combinations can be streamlined into something more manageable.

A trial I was very interested in was the STAND trial looking at the novel combination pretomanid, moxifloxacin and pyrazinamide to treat both drug-sensitive and drug-resistant TB in as little as four months. This combination, known as PaMZ, showed enough promise in phase II trials to move into a larger phase III trial involving 1,500 patients. Should it work, the number of pills required to cure MDR-TB could be cut from 1,400 to 360, all without any unpleasant injections whatsoever, and costing a fraction of the price of traditional MDR regimens. Great, I thought, this sounds brilliant! I should definitely write about that. And then three of the trial's participants died as a result of pretomanid side effects. The trial was put on hold, and while it is due to start again soon, it hadn't at the time of writing this chapter.

So my plans to talk about game-changing, simplified regimens all now rest on the NIX-TB (New Investigational Drugs for XDR-TB) trial, another phase III study brought to you by the organisation behind PaMZ – TB Alliance. TB Alliance is a non-profit organisation dedicated to solving the problem of long and complex TB regimens. They're one of several TB-focused organisations who champion promising science and help turn it into a clinical reality. Their aim is to introduce an ultra-short, simple and affordable regimen that can be used in everyone. NIX-TB is looking at a three-drug regimen consisting of bedaquiline, pretomanid and linezolid to treat XDR-TB in just six months. The best part of this is that all three of these drugs are new agents – bedaquiline has only conditional approval, pretomanid is entirely new and linezolid is only used off-label – meaning that resistance should be minimal. Plus, all of them are oral agents, so no horrible injections. According to Francesca Conradie speaking at the World Conference on Lung Health in Liverpool, every single patient (30, so far) who completed the six-month regimen appeared to have been cured. With more data, the team should be able to apply for regulatory approval and start prescribing the NIX regimen outside a clinical trial. Fingers crossed.

The way clinical trials work is that you basically need to test every combination of drugs in every population of people (for example, drug-sensitive and drug-resistant infections in HIV-negative and HIV-positive individuals, including both paediatric patients and adults). There are currently trials ongoing to look at bedaquiline added to the standard MDR-TB treatment, or delamanid plus the standard MDR-TB treatment. What the NIX-TB trial does is bypass this step by testing something entirely new that could potentially replace the standard regimen. This is surely the future of TB treatment. Ideally, with enough brand new agents, we could start again from scratch with a standard regimen for all TB infections without the worry of whether they are drug-resistant or not. But right now there aren't enough companion drugs for the new agents such as bedaquiline and delamanid, putting them at risk of being added to already failing regimens. This misuse won't cure the patients involved, but it will help *M. tuberculosis* on its way to developing resistance. Will these two drugs go the way that all the others are heading, or is it really possible to preserve them through careful usage? I don't want to sound pessimistic, but without companion drugs, I don't think bedaquiline and delamanid stand much of a chance. For the same reason, a standard regimen is likely an unrealistic dream right now.

Back in 2010, a 35-year-old refugee arrived in Switzerland having travelled from Tibet. Within a month, he was in a Swiss hospital complaining of weight loss, night sweats and a cough that had lasted five months. He'd not previously been treated for TB, but drug susceptibility testing revealed that he was infected with a strain of *M. tuberculosis* resistant to seven drugs – isoniazid, rifampicin, pyrazinamide, ethionamide, linezolid, moxifloxacin and streptomycin. The absence of resistance to an injectable such as kanamycin meant his infection was classed as pre-XDR-TB, but all the same, things were looking pretty crappy for him. He was put on treatment using ethambutol, PAS, capreomycin and cycloserine. Bedaquiline was added in September 2011 and continued until February 2012 (more than six months is not recommended). Soon after, the patient's TB seemed like it was cured. Sure, he had cavities remaining in his lungs but they were thought to be post-inflammatory scarring. The rest of the drugs were stopped in March 2013 after a total treatment time of 24 months. Woohoo! Only, five months later, he suffered a relapse and returned to the hospital with fever and cough.

Second time round, the individualised treatment comprised ethambutol, PAS, capreomycin, clofazimine, cycloserine and amikacin. But then genetic analysis revealed that the strain had accumulated resistance to bedaquiline and, as a result, cross-resistance to clofazimine. Antibiotic therapy was adapted to include meropenem and clavulanate. The strain developed resistance to injectable aminoglycosides and capreomycin. Therapy was adapted again to include levofloxacin, sulfamethoxazole/trimethoprim and delamanid on a compassionate basis. It still wasn't enough. The patient developed severe depression due to the isolation involved in his treatment, and CT scans of his chest showed that the infection was getting worse, not better. The more drastic decision was taken to remove the infected parts of his lungs. It almost killed him, but by February 2015 he was well enough to leave hospital while continuing treatment. Like Phumeza Tisile and many others, he went on to develop hearing loss as a side effect, but he lived. What's important to note here is that he was observed to take every single dose of his medication, and this was a regimen designed by a team of experts blessed with the perks of working in a high-income country. And still, the patient developed resistance. The way that I view antibiotic resistance is as something that is inevitable, but with careful antibiotic stewardship and a constant pace of drug development, we can remain one step ahead of the bacteria. At the moment, though, *M. tuberculosis* seems to be taking two steps for everyone one of ours.

There's so much to be optimistic about when it comes to TB control. All of it is built on a research field filled with a wealth of exciting science. But while the TB drug-discovery field has improved hugely over the last few decades and there is now a real push to find new potential candidate drugs, it's simply taking too long to scale up the use of existing ones to treat all the people who need them. What's the point of new drugs if we're not even using the current treatments to their fullest potential? Take bedaquiline, for example. It can't be used in all the people who could benefit because there aren't enough companion drugs to make up the regimens. Linezolid, in particular, is a key player in the rollout of bedaquiline, but it's not available everywhere that needs it and has toxicity issues. Plus, the use of delamanid in the field is still lagging behind that of bedaquiline, even though we desperately need both drugs to be available together Right Now. A trial testing the two drugs in combination along with others is due to start enrolling patients soon – it amazes me that it's taken so long.

This is one of the biggest failings of the current TB pipeline. It took three years for the phase III trial of bedaquiline to begin following its conditional approval and the drug's sponsor Janssen has been accused of unnecessarily delaying much-needed studies. In addition, delamanid's manufacturers have dragged their heels in providing the drug to patients outside of clinical trials and have only applied for drug registration in a few countries. Koen Andries, however, was quick to point out that there are some huge barriers to getting permission to carry out clinical trials. The blame for all the delays does not rest solely with the pharmaceutical companies involved. They are the good guys in this story, bringing us brand new TB drugs capable of saving countless lives. That's the important part. Without companies like Janssen and people like Andries and his team, we would be in a worse place than we are now. It's also important to remember that, without a real political push, the process of turning the science into reality is going to remain far too slow to keep up with drug resistance, never mind eradicate TB.

What we need is a game changer to bring the advantage back to us humans. Something bigger than new drugs to replace the ones that we're losing through resistance, or even new regimens. This is where host-directed therapies might be able to help in the not-too-distant future. It's a fashionable area at the moment – the idea of targeting the host instead of the bacteria. A bacterium can't develop resistance to a drug that doesn't act directly on the pathogen. Plus, host-directed therapies have particular relevance to the TB field due to how *M. tuberculosis* finds ways to get around the immune system, converting the granuloma into a personalised little fortress in which it both avoids host-mediated killing and simultaneously hides from antibiotics. TB effectively turns the human immune response into something that benefits the bacteria as much as, if not more than, it benefits the host. So there's a lot of interest in finding a way to either give the immune system a little shove in the right direction or to limit the damage caused by the infection.

Host-directed therapies (HDTs) are the culmination of a thousand different threads coming together. As we pick apart the complex layers of *M. tuberculosis*'s hold over the human immune system, we start to see gaps where a new therapy could make a difference. I believe that this area has the potential to solve some of the big problems in TB treatment – long treatment timeframes, in particular, as well as poor outcomes for the treatment of drug-resistant infection and difficulties in curing latent TB. It's not simple, though.

Our immune response to TB is the combined result of our nutritional state, the roles played by other species living in our bodies, memories of pathogens we've encountered in the past or present and our own genes. Everything combines to create a temporal, dynamic disease state that is so much more than a one-on-one battle between man and microbe. Immune factors that play a vital role at one point of the infection can turn on the host at a later stage. Pathways important in one person may not play an identical role in another. All this means that HDTs may need to take into account not just the timeframe of disease but an individual's 'immune signature' and the impact it can have on treatment. It feeds into the idea of personalised medicine, which is a brilliant concept from an individual perspective but not quite so attractive from a cost or practical point of view.

One HDT among those with the most potential to become real options for TB treatment is imatinib. A drug with such an impossible-to-pronounce name could only come from one field: cancer therapy. Imatinib is a tyrosine kinase inhibitor used to treat chronic myeloid leukaemia (CML). In CML, a signalling protein within the cancer cells becomes switched on permanently and the cells run away with themselves. Divide! Grow! Conquer! Imatinib, however, switches the troublemaking protein back off, and the cancer cells die as a result. But what has cancer got to do with TB? It just so happens that M. tuberculosis hijacks this same signalling protein to avoid being killed by the macrophage. Imatinib can inactivate these hijacked signalling pathways, meaning that M. tuberculosis's message doesn't get through. The immune system can get on with fighting the infection without as much interference from the bacteria. The drug works in mice, improving the clearance of bacteria from their bodies even in the absence of antibiotics. Its effects include promoting phagosome maturation and acidification (which helps to kill cells hiding inside macrophages) and autophagy (in which infected cells kill themselves). One plus-point is that imatinib is already licensed for use in humans, so providing everything goes to plan (and there are a lot of hurdles yet to overcome), we could start to see it being used for TB treatment in as little as five years. There are questions, though, over whether such drugs that improve the immune response to intracellular TB will work fast enough, and potently enough, to make a real difference.

Then there's the fact that our current definition of 'cured' isn't always the end of the story for a TB patient. Today, we tend to focus on killing the bacteria in a patient's lungs as the number one aim of TB therapy. Job done, go home. Only, the infection may be gone but the after-effects of TB can linger. *M. tuberculosis* relies on inflammation and the breakdown of lung tissue for transmission of the infection. It's this damage to the lung – and the resulting respiratory failure – that most commonly kills a TB patient. But what I had never considered before is that this damage doesn't just go away when the bacterial infection is sterilised. Something like half of all patients are left with permanent lung damage that can lead to lifelong coughing and breathlessness, not to mention a shorter lifespan than someone who's never had TB. In 1989, a paper followed up on TB patients a whole decade after they'd been 'cured'. They found that 68 per cent of those patients had long-term airway problems the severity of which correlated with the severity of their original TB infection. More recently, studies demonstrated that treated TB leaves someone far more likely to die of a lung-related cause – infections such as pneumonia can take hold far more easily in a TB-damaged lung. When I talked to Robert Wallis of the Aurum Institute, South Africa, he referred to this sidelining of the lung in TB treatment as 'short-changing the patient'. It's like everyone has focused so closely on the microbe that the damage left in its wake has been overlooked.

Using corticosteroids to reduce the inflammation associated with TB was first proposed in the 1950s. It's that old paradox again – we need the inflammatory branches of the immune system to fight off TB, but once you have the disease, inflammation is counterproductive for the host. Robert Wallis has been heavily involved in this idea of switching off the inflammatory pathways that cause the problems in TB. Because steroids are associated with some dodgy adverse effects, interest is now focused on phosphodiesterase inhibitors that modulate the same pathways as steroids. Wallis is about to begin a phase II trial in around 200 patients looking at a compound currently known as CC-11050. It's an anti-inflammatory drug that, in mice, can speed up drug clearance of *M. tuberculosis* and reduce granuloma size and number. So fingers crossed it will have a similar effect in reducing treatment times and lung damage in humans. We will have to wait and see what happens. Like much of the work going on in the TB field, everything in the HDT area teeters on the edge of something tangible with real benefits to real people. A big shining hope for the future that – I hope – proves to be the transformative advance we are all waiting for.

TB Continued

As buds give rise by growth to fresh buds, and these, if vigorous, branch out and overtop on all sides many a feebler branch, so by generation I believe it has been with the great Tree of Life, which fills with its dead and broken branches the crust of the earth, and covers the surface with its ever-branching and beautiful ramifications.

Charles Darwin certainly had a way with words – and with science as well, to give the man full credit. But when it comes to his 'Tree of Life' metaphor for the evolution of all species on Earth, he wasn't being strictly correct. I'd say it's more like a large bush, or a thicket. You know the impenetrable forest of tangled thorns that surrounded poor Sleeping Beauty during her magical coma years? Twisting branches and knotted creepers, and thorns that impale all but the bravest of princes? Yeah, that's a better way of thinking of evolution. Sure, 'Thicket of Life' doesn't have quite the same ring to it, but it's more scientifically accurate. Evolution is basically a big tangled web in which everyone and everything is presumably within six degrees of Kevin Bacon, be they microbe or man. I started out writing this book with the basic premise that the evolution of *M. tuberculosis* and *Homo sapiens* are two intertwined branches twisting together through time. But, like Darwin, I sacrificed scientific accuracy for a nice metaphor. In reality, it's really flipping complicated. You know that poem, *No Man is an Island?* John Donne meant it to mean that no human is self-sufficient; that we need others, and one person's actions will impact upon someone else. I'd like to extend it to say that, physically speaking, Man *is* an island, but one that is inhabited by countless other species, either transiently or permanently. All of these species have grown up in a world where interaction with other species is a way of life and everything has left its mark on everything else.

For TB, this story of mutualism, compromise and outright warfare started long before the first humans embarked on their hunter-gatherer ways in Africa's Cradle of Life. But it was only when people entered into the picture that things started to get interesting. Together with those early-you's and early-me's, *M. tuberculosis* spread to nearly every populated place on this planet. We didn't travel alone. Countless other species joined us from time to time, elbowing each other in the ribs as they staked their claim on their little spot in the universe. Worms, viruses, bacteria, humans and animals. Thousands of stories all knotted together, with humankind and *M. tuberculosis* single threads in the plait. It was against this crowded backdrop of life and death that an uneasy equilibrium of sorts was reached between patient pathogen and pertinacious primate. I suppose that could have been the end of the story. TB could have remained just another disease. Nothing special among all those other species whose histories are wrapped up with our own. But we humans are the catalysts of our own destruction when it comes to infectious disease. Faster than evolution could keep up, human progression marched us towards a new world in which *M. tuberculosis* became more than just a survivor, its success no longer only determined by the DNA of the bacterium and its host alone. Our overcrowded and polluted living conditions, poor diets and social inequalities gave TB what it needed to flourish and become the world's biggest infectious killer. Even today, the story repeats itself again and again with different players and different settings. Every time, it's social factors that provide the sunlight and water for TB's creeping vines to grow.

Science has been trying to cut TB back for centuries. We've gradually come to understand what has made *M. tuberculosis* what it is today, and I can't help but feel awed at just how complex and clever the bacterium is. It's only fairly recently that we've realised latent TB is far from latent – there's a whole world inside a TB patient, inside each granuloma. One population of bacteria might use their thick cell walls to commandeer macrophages as their own, while others swim around in their dinner of host-derived cholesterol. Sometimes *M. tuberculosis* might secrete virulence factors into the host cell to encourage a more hospitable environment, while other cells adopt a silent-running approach and effectively switch everything off until times are better. Maybe the bacterium will turn on efflux pumps to detoxify its innards, or maybe it will survive through quirks of protein expression or growth rate.

All of this science and more feeds into the same question – how do we kill *M. tuberculosis*? Can we develop drugs that interfere with some of the bacterium's vital pathways? Or can we give the immune system a helping hand through vaccination or host-directed therapy, allowing it to unravel millennia of co-evolution and that uneasy equilibrium between man and microbe? The last few years have seen two new drugs gain regulatory approval, with a few other potentials on their way. The vaccine field is looking better than ever, and the rollout of the GeneXpert diagnostic has made me believe that we really can achieve amazing things. The science is all looking pretty good, if you ask me, but it's not moving fast enough. It's a new world out there, with HIV infection, drug resistance, co-morbidities such as diabetes, immune-suppressive treatments for other conditions, malnutrition and no end in sight for those social inequalities that have plagued us for … well, forever. Like a weed, TB continues to slip through the cracks. TB isn't a single problem but a collection of seperate epidemics, each of them driven by a collection of different factors specific to each setting. It follows that to eradicate TB we need to focus on the differences at every level: different locations, different populations, different risk factors.

It's becoming increasingly obvious to me that infection and disease are seperate things. A disease state is as much to do with the host response as it is to do with the pathogen, and how a host responds to an infection isn't solely a product of that individual's DNA. It's the sum of a thousand parts, some of them genetic but many of them down to the lives we live. To me it follows that treating a disease can't focus only on the infectious agent itself, but must consider all the other factors involved in determining whether someone will remain healthy or not. While science has often focused more on the pathogen than on the people, I am hopeful that the future will bring a more holistic approach to TB control. One where the patients sit at the centre and treatment targets not just the infection in their lungs but the entire human – including everything from the microbiome to co-morbidities. And one where curing TB doesn't come with so much collateral damage in terms of permanent lung damage, drug side effects and crushing poverty.

Poverty has been a theme throughout this book. In the introduction, I talked about recasting TB as a modern monster rather than a mothballed curiosity of the past. I have to wonder if the 'TB is a disease of the poor' trope is part of the reason why TB is sometimes overlooked. If TB is simply a barometer of inequality and poverty, then does that mean there's no beating it? Is it truly here to stay until the entire world reaches some futuristic utopia in which everyone has access to up-to-date healthcare and nobody struggles to feed their family? I keep comparing TB to HIV. Over the past few decades, HIV infection has been reduced from a death sentence to a chronic, controllable condition. The HIV advocacy field has succeeded in maintaining momentum when it comes to efforts to return the world to a pre-AIDS state. But then HIV, unlike TB, only recently jumped over into the human population. Not only is the infection right on our own doorsteps, making it impossible to ignore, but it was relatively easy to see ways of helping to bring it under control using cost-effective interventions. TB, in comparison, has been with us forever. While I think everyone can agree that TB is awful, it's not new-awful. Unlike HIV or the more recent Zika and Ebola outbreaks, TB isn't surprising or shocking, or all that close to home for many of us. And it's not going to be fixed without sustained, long-term effort and funding. While the worldwide burden of TB is going down, I can't see how we are going to get close to those aspirational-but-not-very-realistic 2035 targets without something dramatic happening to change the status quo. Much of the problem here is advocacy – and I believe we can learn from the HIV field on this front. We need to find a way to keep TB on the radar and make sure it doesn't drift out of the political agenda.

It's no coincidence that the success of the HIV field has been accompanied by levels of funding 10 times higher than that injected into TB research and development. If we look at how funding levels compare to disease burden, there are huge disparities. For the 2014–2016 period, the Global Fund that controls much of the funding going into HIV/AIDS, TB and malaria in the developing world divided up donations to send 50 per cent into HIV, 32 per cent into malaria and 18 per cent into TB. In 2014, there were 1.2 million AIDS deaths, 0.5 million malaria deaths and 1.5 million TB deaths. One-third of HIV deaths were from TB. A recent announcement from the Global Fund revealed that international donors have pledged nearly $13 billion to fight these three diseases in the developing world over the next three years. It's great news, but I wish that a little more than 18 per cent would go to TB. I'd like to see governments – from both high- and low-TB-incidence countries – committing to long-term investments into worldwide TB control. It's dangerous to believe that because TB is rare in high-income countries such as the UK or US that it's not our problem. Drug-resistant TB is only a plane ride away, after all. Only with everyone working together will real progress be made, and it's heartening to see how the TB world has embraced this collaborative approach. The scientific community has a big part to play in TB advocacy, both in standing up for their field and also by doing the sort of science that helps funders believe that their investment really could make a big difference. I will admit that 'Give us more money' is a difficult argument to make, though, as TB already gets a big chunk of the pot (for comparison's sake, I'll mention that non-communicable diseases such as cardiovascular disease, diabetes and chronic lung diseases kill 36 million people a year – most of these deaths occurring in low- and middle-income countries – yet receive far less funding).

But I don't want to end by talking about money and death tolls. This book isn't about the complicated problem of global resources and cost-effectiveness of health interventions. It's about people and compassion for everyone living on our small, interconnected planet. Sometimes I think this part gets lost among all the science. I know that it did for me back when I was working in the lab. Somewhere among all those gene knockouts and protein purifications and rejected papers, I forgot the real reason why I got into research in the first place. The one thing writing this book has taught me is that there's a lot more to TB than science. Eradicating the disease isn't going to happen as a result of someone performing a successful experiment in a brightly lit lab. No new drug or clever piece of diagnostic kit or even a vaccine is going to fix the problem. It's going to take political will and bold policies that both ensure everyone has access to the current tools and work on the issue of social protection. Until we see universal improvements in living conditions, access to healthcare, the effects of poverty and all the other markers of inequality, then TB is going nowhere. One day, perhaps TB really will be a curiosity of the past, deserving of its old-fashioned image. For now, though, it's very much a disease of the present and, sadly, the future.

I'm not sure that this is a more optimistic way to end a book, but then this isn't the end of the story. There's still a big part left to be written, and I'm excited to find out what's coming next.

Acknowledgements

More than a decade ago, two scientists sat in a dodgy London pub and decided to write a popular science book on all the ways that bacteria are brilliant. Over the years, they plotted and planned and drank quite a lot of wine in the process, but never actually wrote the book. Today, it feels strange to see my own name in print without Suzie Hingley-Wilson's keeping it company. If it wasn't for her, I would never have discovered that I prefer writing to research, and I certainly couldn't have finished this book without all her suggestions and input. One day, I hope that we'll finally write our bacteria book but, for now, Suzie will have to make do with this dedication.

There's nothing like writing about your own field of scientific research to make you realise how little you actually know. I am hugely grateful to all the experts who have given up their time to provide me with insight and direction, and the occasional reality check. In no particular order, they are: Koen Andries, William Bishai, Frank Bromacher, Catherine Cosgrove, Anna Coussens, Thomas Dick, David Dowdy, Lukas Fenner, Helen Fletcher, JoAnne Flynn, Sarah Fortune, Andrew Noymer, Onn Min Kon, Kim Lewis, Helen McShane, Nerges Mistry, Alexander Pym, David Sherman, Nicholas Bellantoni, Claudia Denkinger, Julie Archer, Adrian Hill, Robert Wallis, Helen Donoghue, Mark Spigelman, Charlotte Roberts, Bruce Rothschild, Sébastien Gagneux, John Taylor, Cristina Gutierrez, Don Walker, Samantha Sampson, Pere-Joan Cardona, Cris Vilaplana, Dany Beste, Tawanda Gumbo, Alistair Story, Mike Mandelbaum, Leonard Amaral, Joanna Breitstein. An extra-special portion of gratitude goes to my non-TB friends who, despite having no vested interest in furthering the TB cause, agreed to provide me with feedback. I don't know why you did it, Hellen Thomson, Nick Brereton, Austin Guest, Volker Behrends and Timothy Simpson, but I'm super-grateful all the same.

There are so many more people who I would have loved to talk to and whose published work has played a huge part in filling these pages. Unfortunately, I think that the lovely people at Bloomsbury would have lost their patience with me had I dragged out my deadline by the additional decade it would have taken to speak to you all. I am sure that working with me tested their patience as it was, so special thanks (and blanket apology) must go to Jim Martin and Anna MacDiarmid. Also, thank you to Myriam Birch for her editing and knack of spotting sentences that simply made no sense, and to all the other people who've played a role in making this book a reality.

Years ago, I helped a friend finish his PhD thesis with hours to spare before the deadline. In his acknowledgements, he thanked me for my 'meticulous proof-reading'. Unfortunately, the thesis was riddled with spelling mistakes and one of his viva examiners – my ex-boss – was quick to point out this backhanded compliment. So I hesitate to thank Roger Buxton for his mentorship and for inspiring me to find the fascinating story behind the research, but ... what the hell, thanks. And thanks to Huw Williams for employing me not once but twice, and for providing the backdrop to where I found my love for latent TB and writing, and where I made many of the friends who have helped me with this book. My last batch of thanks goes to my parents for encouraging my first writing endeavours, even if they were about a man who gave birth to an entire football team of sons; my sister Rebecca for being my first reader and the person whose opinion I will always value over everyone else's; and most importantly Phill, for being the only person to believe in this book without needing to actually read it.

Index

Acid Fast Club 71–72
Afghanistan 124, 155
Africa 11, 42, 43, 85, 116, 155, 206,
  260
AIDS 9, 95, 100, 116–18, 262–63
Amaral, Leonard 249
Americas 42, 44–46, 70, 76
Andries, Koen 239–40, 243, 254–55
Angola 107–108
antibiotics 11, 62
  early antibiotic treatments 169–71
  see drug development
antigens 87–88, 110, 138, 153
  Antigen 85A 87–88, 96
  antigenic variation 96–97
APOPO 218, 220–21
Asia 11, 42, 43, 76
asthma 112, 113
Atlit Yam, Israel 21–26, 30, 32, 34
aurochs 33–34
Australia 35, 42, 179

badgers 8, 35–36
Bangladesh 43, 250
Barnes, David 208–209
Barry, Clifton E. III 130
BCG vaccination 83–87, 90, 95,
  98–99, 110
bedaquiline 117, 196, 239–43, 246,
  247, 251–54, 255
Beijing Lineage 43, 57, 124, 235
Belarus 196
Bellantoni, Nicholas 49–53, 63
Beste, Dany 143
Bigger, Joseph Warwick 160, 165
Bill & Melinda Gates
  Foundation 245
biomarkers 138, 223–25
Bishai, William 113
bison 34, 47
  Bison antiquus 29–30, 37
Black Death 27
Bloch, Hubert 148

bones 17–18, 23, 24, 58
  ancient animal species 30, 32
  Pott disease 32, 45, 55–56, 59
Botswana 118–19
bovids 31–37, 40, 47
bovine tuberculosis 34–37
Brazilian Ministry of Health 66
British Medical Journal 141
British Museum 15, 16–17, 19
Brombacher, Frank 153
Brontë family 60, 71
Brosch, Roland 37, 39
Brown family burials, Rhode Island,
  USA 51
Bunyan, John 56
Bynum, Helen 62
Byron, Lord 61

Cafferkey, Pauline 209
Callahan, Charles 58–59
Calmette, Albert 83
Camus, Albert The Plague 181
cancer 13, 17, 61, 135, 171, 201, 256
  bladder cancer 85
  gastric cancer 112
  oesophageal cancer 112
Canetti, Georges 39
Caribbean 35
cats 7–9
cattle 24, 32–36
Centers for Disease Control and
  Prevention (CDC) 194, 215, 227
Central Asian Lineage 43, 44–45
Chesney, Russell 59
chest X-rays 225–27
China 10, 57, 76, 116, 155–56, 196,
  209
  population 33, 156
cholera 62, 93
cholesterol 141, 151–55, 157, 261
Chopin, Frédéric 60
Collaborative Tuberculosis Strategy
  for England 204, 206

Columbus   45–46, 66
Congo, Democratic Republic of
    the   124, 197
Conradie, Francesca   252
consumption   20, 52, 56, 60–61,
    92–93
Cosgrove, Catherine   199–201, 221,
    229
CT (computerised tomography)
    scanning   19, 54, 135

Daily Mail   7, 35
Dar es Salaam   219
Dartois, Véronique   245
Denkinger, Claudia   224–25, 226,
    229, 232, 234
Dheda, Keertan   224
diabetes   115, 133, 155–57, 190,
    222–23
diagnostic tests   218–21
    biomarkers   223–25
    case-finding approach   236–37
    chest X-rays   225–27
    developing diagnostic
        tests   221–23
    GeneXpert MTB/RIF   227–34,
        261
    genome sequencing   234–36
Diamond, Jared   54
Dick, Thomas   244
diet   19, 24, 53, 59–60, 62
diseases   53–54, 58, 261–62
    Digitised Diseases   54–56
DNA   71, 72–73, 120, 131, 150
    pathogens   17–18, 23–25, 29, 40,
        46, 107, 159, 210
Doctors Without Borders   129, 177,
    179–80, 195
Donoghue, Helen   17, 18–19, 21,
    23, 27
DOTS (Directly Observed Therapy,
    Short-course)   69
Dowdy, David   230–33, 235–37
Doyle, Arthur Conan   93
drug development   169–73, 174–75,
    196, 246–47
    bedaquiline   117, 196, 239–43,
        246, 247, 251–54, 255
    host-directed therapies
        (HDTs)   154, 255–57, 258

penetration of TB lesions   244–45
reducing lung damage   257–58
target-based drug
    development   173–74
TB Drug Accelerator
    Program   245–46
thioridazine   247–49
treating MDR-TB and
    XDR-TB   249–54
drug resistance   184–86
    efflux pumps   188–89
    evolution of bacterial
        strains   189–90
    isolating patients   193–95, 235
    non-adherence   187–88
    standardised treatment   186–87,
        190
drug-resistant TB   9, 10, 11, 63, 81,
    119, 139, 175, 196
    see MDR-TB; XDR-TB
Dumas, Alexandre   60

Earl, Ashlee   184
East Asian Lineage   43, 76
Ebola   53, 209, 211, 262
Edmondstone, Sir Archibald   13, 19
efflux pumps 168, 174–75, 188–89,
    261
elephants   31, 32, 36, 47
End TB Strategy   10, 100, 156
epidemics of tuberculosis   60–61,
    66–70, 75–81
    epidemiological tracking   209–10
Ethiopia   110–11
Euro-American Lineage   43, 44, 46,
    76
Europe   35, 42, 43, 70, 125, 179, 197
evolution   259–60
Express   8
extra-pulmonary TB   11

farming   53–54, 57
Fenner, Lukas   213–14
FIND   224, 229
Find&Treat   202, 204
Fitzgerald, Michael   35
Fleming, Alexander   169, 184–85
Fletcher, Helen   89–90
Flynn, JoAnne   90, 135–38, 167–68
Fortune, Sarah   90, 97, 166, 167

free radicals 145, 150–51
Friedland, Gerald 181–83
Friedland, John 130–31
Friedman, Misha 125

Gagneux, Sébastien 40, 46, 77, 96
Galili, Ehud 21, 23
Gambia 111
Gauguin, Paul 60
genetic susceptibility 69–74
GeneXpert MTB/RIF 227–34, 262
genomes 73–74, 111
    genome sequencing 184, 189,
        213–14, 234–36
    genomics 171–72, 242
    GWAS (genome-wide association
        studies) 72–73, 81
GlaxoSmithKline 173–74
Global Fund 263
Goetz, Thomas 93
granulomas 125–27, 130–39, 151–52,
    157, 167–68, 189, 245, 255, 258,
    260–61
Granville, Augustus Bozzi 13–20, 26
Graunt, John 56
Guérin, Camille 83
Guilhot, Christophe 146
Guinea-Bissau 108–109
gut microbiome 112–13, 115, 120
Gutierrez, Cristina 38–39

hearing loss 177–78, 253–54
Helicobacter pylori 111–12, 120, 240
helminthic infestation 109–12, 120
Hernández Pando, Rogelio 134
Hill, Adrian 73–74
HIV 9, 27, 92, 100, 109, 124–25, 186,
    195, 200, 218, 226, 236, 261–63
    immune reconstitution
        inflammatory syndrome
        (IRIS) 95–96
    stigma 102–103
    TB infection 74–75, 77–81, 116–
        20, 138, 155, 181–83, 218, 221–
        22, 224, 230
    vaccination 86, 88
Hobby, Gladys 160
homeless populations 125, 200, 203,
    214–15
hominids 38, 40, 47

Homo 38
    erectus 24, 38, 40
    habilis 38
Homo sapiens 7, 32, 38
    migration out of Africa 41–44
Horn of Africa 38–39
hypoxia 127, 130–32, 137, 168, 175,
    245

ICL (isocitrate lyase) 149–51
immigrant populations 200–207,
    215–16
immunity 12, 67–69, 77, 81
    antigens 87–88, 96–97, 110, 138,
        153
    cholesterol 152–53
    immune reconstitution
        inflammatory syndrome
        (IRIS) 95–96
    immune system 91–92, 97–98,
        112–13
India 10, 43, 84, 116, 123, 155, 186,
    190–93, 200, 205–206, 229, 232,
    236–37
    population 33
Indo-Oceanic Lineage 42–43, 76
Industrial Revolution 57, 58, 62, 75,
    81
infection 10, 11–12, 41, 74–81,
    261–62
    reinfection 11–12, 119–20
    transmission of disease 208–13
    see latent infection
influenza 34, 50, 65, 89–90, 93,
    103–104, 109, 120
    Spanish flu pandemic 104–107
Iraq 197
Ireland, Republic of 35

Janssen Pharmaceutica 239–40, 254
Janssen, Paul 243
Jones, Steve 54
Journal of Infectious Diseases 187

Kafka, Franz 60
Kallmann, Franz 71
Karakousis, Petros 153–54
Keats, John 60
kill curves 160–61
King's Evil 66–67

Koch, Robert   62, 92–95, 228, 240,
        248
Kon, Onn Min   201–202, 229
Kopenawa, Davi   65
Krause, Johannes   46

Lalvani, Ajit   200, 205–206
Lancet, The   116
latent infection   44, 74, 81, 105, 116,
        118, 124–25, 139–40, 260–61
    granulomas   125–27, 130–39, 151–
        52, 157, 167–68, 189, 245, 255,
        258, 260–61
    screening immigrants to the
        UK   205–207, 215–16
Lebanon   123–24
Lee, Chien-Chang   153
Lehmann, Jörgen   170
Lesotho   78
Lewis, Kim   163–65, 167, 174
lineages of tuberculosis   37, 41–42
    Lineage 1   42–43, 76–77
    Lineage 2   42, 43, 46
    Lineage 3   42, 43
    Lineage 4   42, 43, 44–45, 46
    Lineage 5   42, 43
    Lineage 6   42, 43
    Lineage 7   42, 43
    San Francisco, USA   75–76
lipids   18, 23, 24–25, 141, 143–52,
        157, 244
Lipsitch, Marc   68–69
London   57–60, 114–15, 123, 215
    current prevalence of TB   197–99
    drug availability   201
    North London   85, 209–13
    poverty and TB   199–200
    TB amongst homeless   203
    TB amongst immigrant
        populations   200–203
    TB amongst native
        populations   202–204
    XDR-TB   212–13
London Assembly   197, 198, 215
lung microbiome   113–15, 120

malaria   9, 40, 65, 88, 90, 120, 243,
        263
    malaria and TB
        co-infection   107–109

Mandelbaum, Mike   206
Mann, Thomas The Magic
        Mountain   122–23
Mantoux test   95
mastodons   30–31, 47
McConville, Amy   207–208, 211
McKinney, John   162, 165–68
McShane, Helen   87–90, 99–100
MDR-TB   124–25, 182, 184
    prevalence   196
    treatment   177–79
    treatments, developing   249–54
measles   50, 65, 99–100
meningitis   24, 40, 79, 95
microarrays   131–32, 151–52, 166
microbiomes   112–16, 120
microfluidics   165–66
miliary tuberculosis   35–36
Mistry, Nerges   186–87, 190, 192–93
Motsoaledi, Aaron   79
Mozambique   219–20, 231
MRSA   179, 214
Mukamolova, Galina   161–62
mummies   13–21, 26, 28, 46, 107
Museum of London Archaeology   55
MVA85A vaccine   87–90, 99–100
Mycobacterium (MTBC)   37–38, 47
    evolution   39–42
Mycobacterium africanum   37
    bovis   34–37, 83
    canetti   38–41, 46–47
    caprae   34
    microti   34
    mungi   34
    orygis   34
    pinnipedii   34, 46
    smegmatis   38, 239–40, 243
    suricatta   34
Mycobacterium tuberculosis   17, 18–19,
        20–21, 26, 27–28, 46–47,
        260–61
    ancient animal species   29–34
    cholesterol   141, 151–55, 157, 261
    culture   142–43, 172–73, 217, 221,
        226–30
    genome   40–41, 149, 171
    lineages   37, 41–46, 57, 75–77,
        118, 124
    lipids   18, 23, 24–25, 141, 143–52,
        157, 244

Mycobacterium bovis   36–37
persisters 163–68, 174–75
phagosomes   144–48
slow growth   143–44
structure   141–42

National Indian Foundation, Brazil
    (FUNAI)   66
National Institute of Health
    (NIH)   103
Native Americans   45, 66, 77
Natural Trap Cave, Wyoming,
    USA   29, 31, 37
Nature   40, 138, 147
    Nature Communications   54, 244
    Nature Medicine   245
Neanderthals   38
Nejentsev, Sergey   73
Neolithic Revolution   34, 41, 44, 47,
    53–54
Netea, Mihai   98–99
New England Journal of Medicine   80,
    228
Nigeria   10, 210, 211
Noymer, Andrew   106–107

O'Garra, Anne   138
online forums   129–30
Opie, Eugene   133–34
oxygen   127–28, 130–32

Pakistan   10, 111, 200
pasteurisation   34–35
PCR (polymerase chain
    reaction)   17–18, 23, 25, 37,
    213, 228
PDIM (phthiocerol dimycocerosate)
    145–48, 152, 164
Peacock, Sharon   235
persisters   159–60, 163–68, 174–75
PET (positron emission tomography)
    scans   130, 135
phagosomes   144–48, 153, 256
Philippines   43, 76
plague   40, 50 62, 93
pneumonia   79, 98, 257
Poe, Edgar Allan   61
Poland   123
population growth   53–54, 57
Pott disease   32, 45, 55–56, 59

poverty   58–60, 70, 102, 199–200,
    262
Public Health Act UK 1984   235
Public Health England   204, 212,
    215
pulmonary TB   11
Pusch, Carsten   107
Pym, Alex   116–18, 186

Qian Gao   189

Ramakrishnan, Lalita   147
rats, pouched   218–21, 236
Ray family burials, Connecticut,
    USA   50–51
refugees   102, 123–24, 199, 253
reinfection   11–12, 119–20
Reisner, David   71
RFLP (restriction fragment length
    polymorphism) analysis   210
Rhee, Kyu   150–51
Riccardi, Fabio   109
RNA   131, 150, 164
Roberts, Charlotte   25–27
romanticisation   20, 57, 60–61
Rosdahl, Karl-Gustav   170
Rothschild, Bruce   30–32, 37
Royal Institution   13
Russell, David   145
Rwanda   197

Sampson, Sam   162–63
SARS   209
Schmalstieg, Aurelia M.   188
Schneider, Bianca   108
Science   106, 241
Science Translational Medicine   136
scrofula   66–69
Segal, William   148
Sherman, David   146–47, 148
silicosis   79–80
Small, Peter   75–76
Somalia   124, 200
Sousa, Alexandra   66–68, 77
South Africa   10, 11, 77–80, 116,
    179–80, 195, 205, 229, 230
    townships   101–102, 116–17
    XDR-TB   181–84, 193–94
Speaker, Andrew   195–96, 209, 227
Spigelman, Mark   17–18, 21, 26–27

statins   153–54
Stevenson, Robert Louis   60
Story, Alistair   202, 215, 216
Survival   66
Swaziland   78, 231, 233
symptoms   11

Tanzania   214, 218–19
Taylor, John   15–16, 19–20
TB Alert   206, 207
TB bacillus   12, 62
TB Drug Accelerator
      Program   245–46
teeth   25
Temple of the Seven Dolls, Yucatán,
      Mexico   45–46
thioridazine   247–49
Tisile, Phumeza   177–80, 183, 186,
      193–96, 227, 232, 247, 253–54
transmission of disease   208–13
treatment   11–12, 168–69
   early antibiotic
      treatments   169–71
   lung collapse therapy   128–29
   preventive treatment   80
tuberculin   92–95, 228
tuberculosis   7–9
   current death rates   9–10
   current incidence   10–11
   funding research   263–64
   historical death rates   61–62,
      92–93, 105–106, 123
typhoid   40, 93

Udwadia, Zarir   192, 193, 248–49
Uganda   110, 117, 229, 230–32
UK   35, 36, 123, 179, 263
   screening immigrants for
      TB   204–207
   TB control   215
Ukraine   124–25
undernourishment   155
UNICEF   85
urbanisation   24, 43–44, 47, 62, 199,
      214–15, 237
   nineteenth century
      London   57–60

USA   35, 76, 123, 179, 214–15, 263
USS Richard E. Byrd   91–92, 98

vaccination   83
   BCG vaccination   83–87, 90, 95,
      98–99, 110
   developing new vaccines   85–91
   MVA85A vaccine   87–90, 99–100
   tuberculin   92–95
Vaccinia Ankara virus   87
vampires   62–63
   vampire burials   50–53
Van Ginneken, Bram   225–26
Vietnam   43, 76
Vilakazi, Benedict Wallet   78
Villemin, Jean-Antoine   71

Waksman, Selman   169
Walker, Don   55–56
Wallis, Robert   257–58
Walton family burials, Connecticut,
      USA   49–50
websites   129–30
Weetjens, Bart   218
Wellcome Library   129
West Africa 1 Lineage   76–77
Western Pacific   11, 42
Wilkinson, Robert   138
Wilmer, Harry A. Huber the
      Tuber   121–23, 127, 128
Wirth, Thierry   57, 81, 124
World Health Organization
      (WHO)   10, 86, 99–100, 214,
      226, 229, 230, 236, 250
worms, parasitic   109–12, 120

XDR-TB   124–25, 180–81
   London   212–13
   South AFrica   181–84, 193–94
   treatment   177–79, 209
   treatments, developing   249–54

Yanomami people, Amazon
      Basin   65–70, 75, 77

Zika   53, 262
zoonoses   34, 36, 37, 46